GANGWAY!
SEA LANGUAGE COMES ASHORE

JOANNA CARVER COLCORD

WITH A NEW INTRODUCTION BY
PAUL DICKSON

DOVER PUBLICATIONS, INC.
Mineola, New York

Bibliographical Note

Gangway! Sea Language Comes Ashore, first published by Dover Publications, Inc., in 2012, is a slightly altered republication of the work originally published as *Sea Language Comes Ashore* by Cornell Maritime Press, New York, in 1945. The original illustrations have been replaced, and a new Introduction has been specially written by Paul Dickson, for this edition.

4868 6892 7/12

Library of Congress Cataloging-in-Publication Data

Colcord, Joanna C. (Joanna Carver), 1882–1960.
  [Sea language comes ashore.]
  Gangway! : sea language comes ashore / Joanna Carver Colcord ; with a new introduction by Paul Dickson.
     p. cm.
  "Originally published as Sea Language Comes Ashore by Cornell Maritime Press, New York, in 1945."
  ISBN-13: 978-0-486-48223-1
  ISBN-10: 0-486-48223-5
  1. English language—Terms and phrases. 2. Naval art and science—Terminology. I. Title.

PE1689.C65 2012
428.1—dc23

2011044486

Manufactured in the United States by Courier Corporation
48223501
www.doverpublications.com

CONTENTS

Introduction

Captain Lincoln Colcord and his wife, Jane, of Searsport, Maine, departed on the ship *Charlotte A. Littlefield* soon after they were married in 1881. They did not return from their around-the-world voyage until three years later. Joanna Colcord was born at sea in 1882—"somewhere in the South Seas" was about as precise as she could ever be when describing her place of birth. Her brother, Lincoln Colcord, was born in 1883 during a storm off Cape Horn.

Joanna and Lincoln, like other nineteenth-century maritime children, spent most of their childhood and adolescence at sea on various ships. They were educated at sea, and Joanna taught herself photography, which fostered an early interest in chemistry. Her brother Lincoln later wrote about his sister's childhood: "Up to the age of eighteen she spent most of her girlhood at sea on board her father's command, sailing on China voyages; and from this experience she acquired, as if by nature, the essential feeling of ships and the sea. Throughout her youth she lived constantly in an atmosphere of seafaring, in a setting of ocean days, knowing none but the men and women of the sea, seeing nothing but ships and ports about the world, hearing no speech but the nautical vernacular."*

Once ashore, she earned undergraduate and Master's degrees in biochemistry from the University of Maine. She then attended the New York School of Philanthropy. Between 1914 and 1944, Colcord directed the Charity Division for the Russell Sage Foundation and worked for other philanthropic organizations,

---

* Lincoln Colcord's "Introduction" to Joanna C. Colcord, *Songs of American Sailormen* (New York: W. W. Norton, 1938), 20–21.

including the Red Cross. Her many works on social issues include *Broken Homes—A Study of Family Desertion and its Social Treatment,* and *Your Community: Its Provision For Health, Education, Safety, and Welfare,* which was published in three editions.

In 1922 Colcord began collecting American songs of the sea. She later recalled in an interview how this came about: "I went to dinner with some publisher friends, and although I knew nothing of musical notation, came away pledged to begin the research which led to the publication two years later, in 1924, of *Roll and Go,* a collection of American songs of the sea; and which has ever since furnished a hobby and an interesting use for leisure time."

Joanna Colcord originally wanted to publish *Roll and Go* under the name J. C. Colcord in order to obscure the fact that a woman had collected these songs. Her brother wrote to her to argue against the decision: "And now I sit me down to complain of your decision and your arguments as to the way your name is to stand on the title page—neither are worthy of you....you deliberately propose to dodge your sex for obscure psychological reasons. Not a reason you give will hold water for a minute. There is no objection at all to having a woman collect a volume of sea poetry. No reader would raise a question about it for a moment. There is no objection to a woman's having been to sea; it only adds to the interest of the occasion. Here is something new, a real sailor-woman....You have a good right to your sailor knowledge, and I am proud of it whenever I think of you. You ought to glory in it, and make all the capital there is of it." Colcord was convinced, and she published *Roll and Go: Songs of American Sailormen* under her full name.*

Meanwhile, Colcord's voice as a social activist was loud and clear. She was an early advocate of New Deal social reform and a proponent of Social Security, and as war threatened in the late

---

* Letter quoted in Parker Bishop Albee, Jr., *Letters from Sea, 1882–1901: Joanna and Lincoln Colcord's Seafaring Childhood* (Gardiner, Maine: Tilbury House Press, 1999).

1930s, she urged women to prepare to take on industrial jobs to help fight the threat of Nazism. She was also part of a commission created in 1937 by the 20th Century Fund, which argued for the national debt to be cut in half, lest it foster inflation. Roll and Go was revised as Songs of American Sailormen in 1938. In a review, The New York Times observed that until the publication of the original Colcord book, there had been no adequate study of sea chanties. The book had a major impact on the folk revival of the 1950s—the musician Pete Seeger noted in a 2006 television interview that he had met Joanna Colcord at an early age and learned songs directly from her. He added: "She put out a wonderful book, Songs of American Sailormen.... I still sing from it."

Her interest in sea chanties led her to amass a collection of nautical terms. Colcord knew that this was a worthy book idea when she realized how many of the terms she had learned at sea were employed on land as metaphoric speech. The book, Sea Language Comes Ashore, was published in 1945. The Washington Post called it "entertaining and informative" and of particular interest to students, writers, and seamen--amateur and professional. The Christian Science Monitor termed it "a delight to all who enjoy the American language." Colcord missed the age of the personal computer and the Internet, but she no doubt would have been intrigued by their seaworthy allusions, such as "ports" and "portals," "firewalls" and "surges," "navigation bars" and "docking stations," as well as by bulky desktop computers— "boat anchors"—increasingly being given the "old heave ho" or "deep-sixed" by their owners.

The images she took of Asia as a child were of such high quality that they are exhibited and held in the permanent collection of the Penobscot Marine Museum in Searsport, Maine. Some of the images can be seen on a web page that the museum has dedicated to her: http://www.penobscotmarinemuseum.org/photo-collections/joanna-colcord.html. Colcord also is celebrated with a page on the site http://umaine.edu/folklife/exhibits/women-folklorists/joanna-colcord/. A YouTube video has been posted with her images, and her 1909 Master's thesis can be viewed through the URSUS online system and at www.library. umaine.edu.

In 1950, at the age the age of sixty-eight, Joanna Colcord married her longtime friend, Frank John Bruno, and moved with him to St. Louis, Missouri. She died there ten years later.

—Paul Dickson

To the long-vanished ships in which my young years
were spent

Charlotte A. Littlefield
Clara E. McGilvery
Harvard and
State of Maine

and to the brother who shared that childhood

Lincoln Colcord

We Take Our Departure

That is the way a ship's logbook starts off at the outset of a voyage. "We take our departure" from the last land sighted—be it Sandy Hook or Flattery, Sable Island or the Leeuwin, Diamond Head or Table Mountain or Cape Clear. As it drops below the horizon, its bearings are set down, and that is the starting point of the daily reckoning of distances and directions sailed. So the ship's "departure" is not an end, but a beginning.

The book that here takes its **departure,*** is strictly for fun. It might be called a study in semantics, but I was not educated in that field—which is to say that I am "hollering Uncle!" beforehand, if any philologist, etymologist or lexicographer should happen to fall afoul of it. I *do* claim to know the words that people use at sea and alongshore, and what they mean by them; but when it comes to word-derivations, I am out of my latitude.

Landsmen's Borrowings

At a meeting of an editorial staff which I attended not long ago, the editor-in-chief, speaking of inroads made on the staff by the war, said that the magazine had been proceeding under **jury-rig** for some time past. A little later in the discussion the business manager remarked that it was going to be necessary to box the **compass** of reader interest. Both of these men were from the Middle West, and had no connection with the sea.

---

* Throughout the text, **bold face** is used to steer the reader to the proper reference in the list.

The Mayor of New York City in a recent radio address accused some people, whom he didn't seem to care for, of "throwing a **harpoon**" into the interests of the consumer. In the foreword to a new English book on garden cities, I noticed the following nauticisms within a space of five pages:

> It was impossible to trim our **sails**
> He had not only burnt his **boats,** he had burnt ours
> A negligible **crank** (but this one is doubtful; see the
> list, page 213)

These instances are given only to show the wide distribution among users of the English language of terms borrowed from the sea. Our common speech bristles with similar examples, but in using them, most people are not aware of the connection. How many folks who say that they were taken **aback** by an unexpected happening, know that they are likening themselves to a square-rigger caught by a shift of wind from ahead, and with sails pressed back flat against the masts? Or that, when they mention a couple of their friends as being at **loggerheads,** they have placed them in a whaleboat that has just made fast to a whale, according to some authorities.

The extent to which land speech has borrowed from the sea varies, of course, with the distance inland. In the New England seacoast village of fifty to a hundred years ago, most of the people had firsthand contact with seafaring, and fishermen and homecoming sailors contributed directly to the local dialect. (The coastal speech chiefly referred to here is that of Maine, Cape Cod and the whaling ports of Long Island Sound.)

Professor Chase tells of "an exasperated mother chasing her twelve-year-old son in a gale of wind.... The mother, with her wide spread of skirts, was bearing down upon him and fast over-hauling him, when his eight-year-old brother sung out 'Take her on the wind, Jimmy!'" Only a 'longshore boy could have said that.

Unfortunately, most of the old sea terms are dying out along the coast, except, of course, as they have become embedded to stay, in the language of people in general. But where this list says "in common use alongshore" it might often have been more exact to add "fifty years ago." When these expressions are used nowadays, coast dwellers not infrequently remark in surprised recognition, "Oh, I remember hearing my grandfather say that!"

Many terms used in literature, both poetry and prose, refer to the sea, but were never part of the sailor's language. O'er the glad billows, tempest-tossed, the voyage of life, the sport of wind and wave, true as the needle to the pole, and similar flights of fancy, while they draw upon the seaman's experience, were never on his lips, but belong rather to the library and the pulpit. Where they appear in the text, these expressions have been dubbed literary or poetic.

Other terms actually originated by seamen no longer have any relation to seafaring, and have long since vanished from the sailor's tongue. The **antenna** on an insect's head, or on our radio sets, was to the Roman sailor the sprit or yard on which he set his sail. The **galleon,** a definite type of vessel hailing originally from the ports of the Mediterranean, now sails only on the pages of poetry—as in Francis Thompson's lovely lines, containing a wealth of sea-images:

> Oh, may this treasure-galleon of my verse,
> Fraught with its golden passion, oared with cadent
>   rhyme,
> Set with a towering press of fantasies,
> Drop safely down the time,
> Leaving mine isléd self behind it far,
> Soon to be sunk in the abyss of seas,—
> As down the years the splendour voyages
> Of some long-ruined and night-submergéd star.

Scurvy, that terror of the early seafarer, has been borrowed by shore people for purposes of general abuse (like scrofulous, from the early name for tuberculosis), but no sailor of today has ever seen a case of scurvy.

The terminology of trade and transportation on land owes a heavy debt to sea-borne commerce. When the railroads were new, they coined some words for their operations, but borrowed others outright from ships. Not only was **ballast** taken over in this way, but it is probable that the first ballast used on railroad tracks was the sand and gravel which had been dumped by vessels arriving light, before they began to take in cargo. The run—between New York and Chicago, for example—is the ship's run applied to trains. The **stateroom** on a Pullman car comes from old-time passenger vessels—even the conductor's "All **aboard**" is a direct importation from sea language.

Professor Chase writes:

We speak of **shipping** by rail and of shippers and shipments, as well as consignment and consignee, which are also nautical terms. We board a train and are then "on board," which originally referred to the deck of a ship. We have also borrowed **trip, passenger,**° **fare, freight; berth** and **cabin** on a sleeper; tender or the fuel car attached to a locomotive, from **tender,** a supply or service boat accompanying a ship; and **caboose,** which we took from the Dutch, as an alternative term for the galley or kitchen on a ship, and transferred to the car on a freight train reserved for the use of the train crew.

Many other trade terms could be added to the list—**charter,** as it relates to hiring or engaging means of transportation, was originally applied only to ships. The origin of the **fleet** of motor trucks, taxicabs, and so on is obvious. But the newspaperman's **dogwatch,** the baseball player's on **deck** and in the hole **(hold),** the contractor's wrecking **crew,** the bus-driver's and the soda-fountain clerk's double-deckers, are also sea-borrowings. The airplane has naturally taken over many sea terms, including **pilot** and **skipper.**

Perhaps the most interesting of these industrial borrowings is the graveyard shift, applied to night workers in various trades. This comes from the graveyard watch of the modern steamship sailor—his name for the middle watch, from midnight to 4:00 a.m., which Bradford explains as due to the many accidents at night. But this was not the original name! To the old-time sea-

---

° This can hardly be called sea-borrowing, however, since travelers by stagecoach were called passengers, and the term originally meant only a person making a "passage" or journey.

man in sail, the middle watch was not the graveyard, but the **gravy-eye** watch, when the eyes felt sticky from sleep. At some time in the course of the years, by some slip between tongue and ear, the picturesque if somewhat disgusting old name has got changed to the meaningless graveyard.

Landlubbers' Errors

Not all sea terminology is as correctly used by lands-people as in the examples given in the first section. The *New Yorker* for February 12, 1944, says that one Captain Marshall, in 1788, "put ashore" at several of the Marshall Islands. Strictly speaking, this would mean that he had the misfortune to run his ship aground more than once while in the Archipelago; for put ashore must always carry an object—cargo, passengers, or whatever. The author was probably groping for "**put in**," but that would have been wrong, too, for a ship puts in to a port of call only when in some sort of distress. What he should have said was "**called at**" or "**touched at.**"

A prominent official of the United Service Organizations (USO), when asked at a public meeting what they proposed to do for seamen in the **merchant** marine, answered brightly that his organization was setting up large recreation projects for the Marine Corps at Quantico and elsewhere; and it transpired that he had never understood that there was any difference between the two services. Mistaken ideas such as this have been pointed out in the list. For example, under the word **boat** will be found quite a discussion of the types of craft to which it is, and is not, correctly applied. An attempt has been made to straighten out the understandable confusion in the landsman's mind between getting the anchor broken **away** from the ground, and **aweigh;** and getting the ship under **way.**

Rock, as a verb to describe the motion of a ship, has been included simply to point out that it never should be so used. I have "taken a poke" at the shore-writer's notion that a **monsoon** is a violent cyclical storm, likely to overwhelm a vessel; at his "smart-Aleck" use of **windjammer,** and at his rooted belief that **sail** and **sheet** are synonymous, as well as **hatch** and **hold.** Some effort has also been made to point out errors and misconceptions creeping in during the transfer of words from sea to shore— such as the gravy-eye—graveyard mix-up just described. The landsman's "all **plain** sailing" was plane sailing—i.e., by Mercator's Projection instead of by dead **reckoning**—to the mariner of old. The shore substitution of the Devil to pay for the seaman's **Hell to pay** is shown to rest upon the mistaken idea that to **pay** involved the transfer of money. "Not room to swing a cat by the tail" got its last three words from the landsman—actually, no cruelty to animals is implied, but rather cruelty to humans—for the cat is the bo-sun's **cat-o'-nine-tails.**

Which Was Launched First?

A difficulty that must be honestly faced in discussing borrowings from sea language is: Which came first?—a problem that often stumps the etymologists themselves. Did to make a **spread** refer originally to a ship's spread of canvas, or to a well-furnished table? Was the bridge builder who first called the upstream edge of a bridge pier the **cutwater,** thinking of a ship's bow, or did he simply hit upon the same word to describe the same appearance? Was **bowling along** taken from the ancient game of bowls, or was it the other way about? Which was first derived from the name of a small village in France—the martingale of a horse's neck harness, or the **martingale** under a vessel's bowsprit? What relation, if any, exists between **whipping** the raw edges of a seam to prevent fraying, and whipping the end of a rope for the same purpose? And if there is a relation, which was christened from which? When doubt of this nature exists, I have tried to show it in the text.

This difficulty is enormously increased by the fact that only a few words have been actually coined by sailors in modern times. See, for a few examples, **Grog, Lobscouse, Junk, Slush** and **Steeve.** But most of their words have been taken out of the common fund of English speech and adapted to nautical uses. When these words reappeared in shore speech, they had "suffered a sea change." What now stamps them as belonging to the sea is some combination, some peculiar and exact joining of ordinary syllables so as to have a precise maritime significance. Take, for example, the expression "**Greasy** luck!" This is the farewell to a departing whaler—"may you come back with plenty of oil!" People with a whaling background use it "at every **tack** and turn"—it appears on the flyleaf of a book presented to me, for example. On what possible occasion in the lives of shore people could this combination of words have come together?

These shades of meaning are sometimes so slight that only a seaman's ear could recognize them without explanation. See, for example, such phrases as **Back and fill, Cast about,** go by the **board, Hard up, High and dry, Knock off,** know the **ropes,** part **company,** spin a **yarn,** take in **tow, Turn in,** and an eye to **windward.**

Still a further difficulty lies in the fact that only one meaning of a word with multiple definitions may apply to the sea. To **batten** down is a sea term, but to batten upon, i.e., grow fat as a parasite, has no such derivation. Charter, mentioned above, was first used to mean a grant of power or privileges by a sovereign, and it still persists in this sense in relation to banks, universities, and the like. But to **charter** any means of transportation is an inheritance from seafaring.

The Lingo of the Sea

Says Russell: "Sailors' talk is a dialect as distinct from ordinary English as Hindustani is, or Chinese. English words are used, but their significance is utterly remote from the meaning that they have in shore parlance."

This is an exaggeration; but it is true that sea language is a sub-dialect of English, that it is common to both American and British seamen, and that it is subject to far less variation than exists between the various English-speaking countries, or even between different sections of the same country. An Australian sailor, for example, can go on board an American vessel and instantly comprehend all the orders given; whereas in the continental United States, he would be continually puzzled to know just what was meant by what was said. Americans from both sides of the Mason and Dixon Line speak the same jargon when serving in ships.

Sea English is an old dialect, and even though the sailing ships in which it originated have largely vanished, the terms have persisted. It has been interesting to compare the *Universal Dictionary of the Marine* issued in 1769 by William Falconer (who was lost in that same year when the *Aurora* frigate, in which he served as purser, went down with all hands off the Cape of Good Hope) and the *Glossary of Sea Terms* published in 1942 by Gershom Bradford of the U.S. Navy; and to note the large proportion of words and phrases that are precisely the same, in spite of the lapse of nearly 175 years between publication dates.

The majority of English nautical terms are descended from the Anglo-Saxon and Norman or borrowed from other Teutonic languages, particularly from the Dutch, England's close neighbors and rivals upon the sea. See **Aloof, Boom, Cruise, Deck,** and many others throughout the list.

Many terms of Latin origin were borrowed from the French across the Channel. See **Captain, Large, Prow, Rummage,** etc. **Anchor,** the Latin ancora, seems to have made its way directly into the Teutonic tongues without being first naturalized into French, and was possibly used by the Angles and Saxons before they invaded England. The **compass,** although its present-day name is made up from two Latin elements, was not originally called thus by English seamen. Like other terms relating to navigation by instruments, it is part of the scientific vocabulary devised by early European scholars.

Spanish and Portuguese contributed a small but interesting group of words, and the Far East through its products and the native corruption of English words (pidgin English) added more.

Many tags and phrases used by seamen and familiar to coastal dwellers come from stories which are no longer remembered. The "nub" remains, but the tale is no more. Nobody recalls who was that early exemplar of ca' canny, **Tom Cox,** nor the **Barney** who kept his brig in such a terrible state. We don't know the name of the Admiral who was **tapped,** or in what respects

Charley Noble resembled a stovepipe; nor does memory state what, precisely, was **Jackson's** difficulty—though I suspect that it would be well not to inquire too closely into that.

I make no apologies for the vulgarisms appearing here and there in the list, for they are part of the picture. It must not be imagined, however, that bad language was a commonplace to children on shipboard. Seamen are very modest—even prudish—in what they are willing to let children hear, and women whom they regard as ladies. I was a girl of twelve or thirteen when one day, as we were lying alongside the quay in Newcastle, New South Wales, I heard a teamster talking to his horses. I ran down into the cabin with a new phrase—its cadences were pleasing to my ear, and I had never heard it in all my years of seagoing. "What is a son-of-a-bitch?" I asked.

A few years ago, I went to Sailors' Snug Harbor to see my old friend "Jimmy Star"—God rest his soul, he's gone now; to Fiddler's Green, I hope—and found him in the Harbor's hospital with a busted ankle. "How'd you come by it, Jimmy?" I wanted to know. He turned all colors, gulped a couple of times; and finally brought out: "I wouldn't lie to a lady like you; I was drunk." And that's what sailors are like. Their Nice Nellyism has been a handicap, in collecting material for this book and for an earlier one; but in the course of a long life, I have gathered some samples of what is known in academic circles as "unconventional English"; and when it seemed tolerably certain that one of these expressions originated at sea, I have put it in the list. It is a fair inference that it must have obtained currency in coastal communities almost as soon as it was minted.

It is strange that some useful sea words did not get into circulation ashore—for example, "to bream," which meant to burn off the weeds and dirt from a ship's bottom with torches after **careening.** That describes fairly well the operation of plucking and singeing a fowl; but it was never to my knowledge so used. Terms employed in navigation by instruments, especially resist transplantation. To be sure, **reckoning, latitude** and **departure** have come back after long sea voyages bringing with them into the language of the land the smell of the salt spray; but other navigational terms, such as variation, amplitude, deviation, are so tough and precise in texture as to defy even figurative use ashore. For this reason, a great many expressions in daily use at sea are not even mentioned in the list—they name things and actions so indigenous to ships that even in coastal communities, it is impossible to twist and turn them to advantage.

Cargoes from Abroad

If to this list of words had been added the names of all the foreign goods that sailors first introduced on this continent, it would have included most of the imported luxuries of our early years, except the European goods already known to the settlers when they arrived. Take the cabinet woods with which ship-owners and masters embellished and furnished their mansions—mahogany, rosewood, and ebony; camphor, sandal, and olive woods and many others. These were not in the beginning articles of commerce in this country, but materials discovered by seafarers and brought home stick by stick to fill definite needs. I still have the bill for logs of Santo Domingo mahogany which my grandfather purchased and carried home to be the stair railings and newel post of his new house. Some of these woods went back to sea again, as the wainscoting of the handsome **bright**work cabins of the big sailers.

Or take foods—sugar, rum and tamarinds from the West Indies; tea and preserved fruits from China; spices and coffee from the Dutch islands of the East Indies; curry powder and chutney from India. Or fabrics—Chinese brocades and embroideries; grass-cloth from the Philippines; calico, muslin, madras, nankeen, and the many silks—bengaline, pongee, shantung, surah, tussore—all named from the places where they were first acquired by sailors and brought home as presents to their womenfolk. To include many of these would have stretched the list to an unconscionable length.

The Bill of Lading

The purpose of this little book is to bring together as complete a record as possible of words and phrases developed at sea which have thereafter had some currency upon the land, either in a literal sense (as describing objects and processes closely related to those employed at sea) or figuratively and with completely different application. Words are not included which mean precisely the same thing at sea and on the land unless they originated in man's use of wind and tide, ships and the ocean.

This purpose has been extended somewhat by the inclusion of detailed material on the proper and improper use of certain words in their nautical significance—see, for example, **Forms of address, In** (2), **Put** in. This is chiefly because the subject interests me—it may or may not interest the reader!

Some words and phrases are included, because they always appear in such lists and are commonly believed to have a nautical origin; but when competent authorities believe this to be untrue, I have been at pains to indicate it. As a further safeguard, there will be found at the back a list headed "Doubtfuls."

I have tried to avoid unusual directions to the reader; the two employed because of their convenience being cf., meaning compare with some shore term not listed, and q.v., which means turn to this entry in the list. A term printed in **bold face** in the text always means a cross-reference.

We Dip Our Colors!

This, then, is no comprehensive dictionary of sea terms. Many such books are available, but those found most helpful for this job, in addition to Falconer's and Bradford's, earlier mentioned, have been Admiral W. H. Smyth's *Sailors' Word-Book* issued in 1867 (the worthy Admiral runs to no less than 744 pages of close print!); W. Clark Russell's *The Sailor's Language*, which came out in 1883; and Frank C. Bowen's *Sea Slang*, which is undated, but appeared about 1929. The etymological dictionary used has been that of Ernest Weekley. Eric Partridge's *Dictionary of Slang* has also been consulted freely.

Very little is available in print, however, upon the more restricted subject which concerns us here. An occasional squib appears in a newspaper or magazine about some peculiar phrases thought to hark back to the sea. The only serious treatment that I have come across, however, has been given by two men. A short article by George S. Wasson appeared in *American Speech* for June, 1929, on "Our Heritage of Old Sea Terms." (Mr. Wasson was not a philologist, but he had the most delicate ear of any New England writer known to me for the nuances of coastal dialect.) Professor George D. Chase, a student of philology, has contributed a comprehensive discussion on "Sea Terms Come Ashore," which appeared as a volume in the University of Maine Studies, Second Series, No. 56 (February, 1942), and in the preparation of which the author of this volume bore a hand. Professor Chase, in turn, has examined this manuscript with great care, and given it the benefit of his criticism.

Hoist the Blue Peter!

Well, now the decks are cleared; it is time to weigh anchor and get the ship under way—and so away she goes.

Joanna Carver Colcord

We could not have the sea,
But we could remember words
Like white sails blowing
Among the wings of gulls...
We could not leave on ships,
But...we could have words.

from "Words Like Sails" by Kay DeBard Hall,
in *Christian Science Monitor*, Feb. 18, 1944.

A 1. First-class; from the rating formerly given to British naval vessels, and to merchant vessels for insurance purposes. Very common in shore speech. For emphasis, one may say "A number one," or for something super-excellent, "A number one and a dot."

Aback. Of a vessel, unmanageable, due to a sudden shift of wind striking the sails from the side opposite that to which they are trimmed. To be taken (flat) aback—that is, to be surprised, disconcerted, flabbergasted—is an expression very common on the lips of landspeople, and is often pronounced "taken back" by those who have no knowledge of the sea.

Abaft. Behind (but on board the vessel; behind and outboard is astern). The word is common in coastal speech: "You'll find my valise up attic just abaft the chimney." See **Aft, Beaft.**

Abeam. At right angles to the length of the ship, and some distance off. Common in coastal dialect: "When you've brought the meetin'house abeam, you'll see the Town Hall round the corner." See **Beam, Wind** (2).

Able seaman. Called an A. B., abbreviation of able-bodied. One who is conversant with all the skills required in a man **before the mast.** See **Ordinary** and **Seaman.**

Aboard. Inside the bulwarks, on deck, or (figuratively) in collision: "She was almost aboard of us before we saw her." In this use, the preferred intensifier is close, often pronounced "clost."

"Come aboard!" is the conventional invitation, without which no seaman would enter another's vessel. On the coast, it is a cordial invitation to a person already at the door; varied by "Welcome aboard!" The sailor prefers aboard to on board in most connections. To have (a cargo) aboard means to be in liquor. "He's never ugly, no matter how much he has aboard." See **Cargo.**

On the land, aboard is used in connection with entering any conveyance. "All aboard!" is the starting signal regularly used by trainmen. See **Board** (3); also **In** (2).

1

About, to come, go. To pass from one **tack** to the other. "About (or 'bout) ship!" and "Ready about!" were orders used in tacking ship. On the coast, the former is used as a jocose warning to avoid colliding with someone. (The land use of "to come about" meaning "to occur, take place," probably has no relation to this sea use. But see **Cast** about.) See also **East** about.

Aboveboard. Above the water line. In shore speech, it means frank, open, fair dealing. (Some etymologists derive this use of the word from gambling, when it was required to keep the hands above the table.) See **Board** (1).

Across (Sea and coastal pronunciation acrost). Probably originally a sea term; same as **athwart.** To run across, to find or meet accidentally, is common in shore speech.

Act of God. An insurance term meaning due to causes beyond human control. It was first used in connection with marine insurance.

Adam was an oakum-boy, since. This phrase comes from the shipyards, where it was a boy's job to keep the caulkers supplied with oakum and other needments. It is fairly common alongshore to mean from time immemorial. The British expand it by adding "in Chatham Dockyard."

Adrift. Untied, floating away. It also means rickety, coming apart. In coastal speech, to break, come, fetch or strike, adrift is to become untied or to come apart. "You can't mail this here passel; it's all adrift." To cut, cast, set or turn, adrift means to abandon. Many of these phrases are found in shore speech. See **Drift.**

Afloat. Clear of the bottom; opposite of **aground.** Also, floating in the water, above the surface. In land speech, to keep afloat is to maintain solvency. A curious land perversion makes all afloat mean full of water, drenched or awash. To set afloat is said principally of rumors. See **Float, Hell afloat.**

Afoul. Entangled, or in **collision.** See **Foul.** Ashore, to be mixed up with. To fall, get or run, afoul of are common in shore speech for meet accidentally, or get into an altercation with.

Aft, After. (1) The hinder end of the ship; the portion behind the mainmast where the officers have their quarters. All hands live aft means there is no discipline.

(2) Behind, but within the vessel; behind and outboard is **astern.** The use is best shown by examples: "The after house is aft of the break of the poop." "Take it aft and heave it over so's it will go well astern." See **Abaft, Beaft, Bring aft.**

Afterguard. The officers; those who live in the after house. In coastal dialect, to put in the afterguard means to promote. See **Bring aft.**

Aground. Stranded. In shore speech, to run aground is to meet with disaster of some sort. See **Ashore, Ground.**

Ahead (of). In front of and outside the ship; intensified to dead ahead. "A seaman's term, used in the United States for every possible forwardness that can be imagined." (Chase). One gets ahead financially; is ahead of the times in one's views, etc. See **Forward.** See also **Forge ahead, Heave** ahead, **Full steam ahead.**

Ahold (Pronounced aholt). Grasped by the hands. To take, or get, ahold means, in coastal dialect, to begin to have effect ("this painkiller takes aholt good"); to hear ("I don't know where she could 'a' got aholt of that yarn!"); to assume the management of ("the store was goin' on the rocks till Ellick took aholt of it"), and similar uses. See **Hold** (2).

Ahoy. (Derived by some authorities from the introductory a plus hoy from Dutch hui, come.) The conventional word in hailing a vessel is "Ship ahoy!" if she is being hailed from a boat or another ship; "Ahoy the deck!" if she is being hailed from the dock. These phrases are used jokingly along the coast.

Air(s). A breath of wind, a zephyr. An entry recurring in every ship's logbook is light airs. See **Catspaw.**

Alee. See **Lee.**

Allee samee. Like. "Allee samee hell you will!" Pidgin English, brought home by sailors.

All hands. See **Hand.**

Alligo Bay. Pronunciation of Algoa Bay, South Africa.

Allowance, on. The situation when provisions or water run short and have to be rationed to the people on board a vessel. It is occasionally used alongshore to characterize conditions of stringency.

Aloft. Above the deck; upon the mast or spars. Its use poetically ashore for in the skies may have been the earlier. In coastal dialect, it means upstairs. "Lay aloft and pitch down some hay."

Alongside (of). Beside, particularly, in the water beside; said of another boat or vessel. Used in the same sense alongshore: "You'll find it right alongside of the back porch."

Aloof (Dutch te loef, to windward). Nearer to the wind. Now obsolete at sea; ashore holding aloof means indifferent in manner or distant.

Alow. Down, downwards; as in the old song about John Paul Jones in the *Ranger:* "As bending alow her bosom in snow, she buried her lee cat-head." The word is now obsolete except in the phrase, alow and aloft, meaning completely, all over.

Amidship(s). In or toward the middle of the ship's lengthwise dimension. Used in the same sense alongshore. "Be sure you get that crossbar nailed square amidships." Of a person, amidships means, of course, his belly. "The cow kicked him square amidships and laid him out stiff as a handspike."

Anchor. (1) From ancora, the only nautical word adopted into the Teutonic languages directly from the Latin. The anchor appears in many figurative phrases in both coastal and land speech. To come to, or ride at, anchor is to settle down; to drag anchor means to slip, lose ground; to drop, or cast, anchor signifies to locate oneself permanently. (This last goes back to earlier phraseology of the sea. The modern seaman does not drop or cast his anchor—he lets it go.) Holding in a race horse so as to throw a race is, in track slang, dropping anchor.

To ride to a single anchor means to have it easy; to swallow the anchor is to give up seafaring and settle ashore. An anchor to windward (sometimes sheet-anchor or kedge-anchor) is the same as a nest egg, something to fall back upon.

In coastal dialect, to go ashore with both anchors on the bows describes an inexcusable lack of competence. "Bring your backside to an anchor" is a hearty, though vulgar invitation to take a seat. See **Dutchman's anchor.**

(2) In addition to its obvious meanings, this word also means, in coastal speech, to place a weight upon something likely to blow away, or to fasten something firmly at the base. "Be sure you anchor that picnic cloth good and solid."

Anchorage. A harbor or roadstead where ships may anchor. A favorite name for the home of a retired seaman.

Anchor ice. Ice that forms on the bottom of a lake or pond.

Anchor watch. Not one of the regular watches (see **Time, Watch**) but a night watch stood by one man when the vessel is at anchor, to see that nothing goes wrong.

It is used alongshore when there is occasion to stand guard at night for some reason.

Antenna. The Latin name for a sprit or yard. In modern scientific Latin it means an insect's "feeler," and it appears in radio terminology as the name for an aerial.

Apart. Originally a sailor's word, from his verb, to **part,** q.v.

Appletree-er. A coaster—or the master of a coaster who never lets her get out of sight of land.

Arab (Pronounced Ay-rab). A reproof to a naughty child: "You little Ay-rab, you!" From the Barbary corsairs, reports of whose savagery were widely current in seafaring communities (just as "you little Sanap" followed the atrocities of the French and Indian wars, and "you little Hessian" the Revolutionary War). This is no doubt the origin of the landsman's street Arab.

Argonaut. A name applied to the California gold-seekers of 1849; from the classic myth of Jason and his crew of the *Argo*, who sought the Golden Fleece.

Argosy. A merchant fleet, originally one that hailed from Ragusa on the Adriatic. Its use in shore speech is poetic and literary: "An argosy of rhyme," etc.

Ark. A freight-boat or scow, especially the Biblical vessel built by Noah. In shore speech, something large and clumsy: "a great ark of a piano." It survives also in the child's toy called Noah's ark.

Arrive. According to Chase, the derivation of this word (ad-ripare, to come to shore) denotes its sea ancestry.

Articles. A written agreement by the crew to make the voyage, signed before a government official before sailing. See **Ship** (4). In coastal dialect, to sign articles is to take employment; or, figuratively, to get married. **As long as she creaks, she holds!** A coastal phrase, used about any object under stress, but referring to masts and cordage under press of sail.

Ashore. To go ashore has two distinct sea-meanings: the ship goes ashore when she strands; her people go ashore by walking down her side ladders. In both these connections, the sailor prefers ashore to on shore. See **Aboard.**

The term is in common literary use, ashore and afloat, etc. "All ashore that's going ashore!" is an order still used on passenger vessels, humorously in other connections, on the land. See **Aground,** go ashore to **windward;** go ashore with both **anchors** on the bows.

Astern. Outside of and behind the ship, intensified to **dead** astern. To drift, drop, fall, go, or run, astern means in coastal dialect to fail of success. "I run astern pretty bad last year when the price of potatoes went down." See **Aft, Beaft.** See also astern the **lighter.**

A-taunto (Pronounced a-tanto). From the French autant, to the full. This originally meant, with masts and spars aloft and fully rigged; by extension, in first-class order, but it is rarely heard nowadays.

Athwart (Pronounced athought). Across; used in the same sense alongshore. "There was a big truck right athwart the garage door." See athwart the **hawse.** Athwartships means at right angles to the keel; alongshore, it means crosswise. "This piece wants to go right 'thwartships of the crate, **so-fashion!**"

Avast. From the Dutch houd vast, hold fast, or Portuguese abasta, enough. An order to cease heaving. See **Belay.** An old term, only occasionally heard, and then in abbreviated form: "'Vast!"

Awash. Floating on or barely above the surface.

Away. In motion, or loose; often confused with **aweigh.** The anchor is broken away, or loosed from the bottom, in the process of getting it **aweigh** and the ship under way, q.v. To cast away means to run a vessel ashore and lose her. See **Castaway.** The order to cut away always means to cut away the masts. See also **Break** away.

Away also indicates the direction from the ship. "Where away?" is the reply to the report of something sighted; "There away" or "Right away—three points on the starboard bow," (or whatever) is the routine reply.

"Away she goes!" is the word when a ship begins to move down the launching-ways. It is sometimes used of other objects alongshore.

Aweigh. Of the anchor, off the bottom; in process of being weighed. Familiar ashore through the popular Naval Academy song, Anchors Aweigh.

Awning. This word first appears in Capt. John Smith's account of his voyage to Virginia in 1624: "We did hang an awning (which is an old saile) to... trees to shadow us from the sunne." Since Captain Smith felt it necessary to make this explanation to his readers, we may safely infer that it was a sea term.

Aye, aye, sir. The prescribed acknowledgment of a superior officer's order; used jocularly in other connections ashore. An old joke was: "Who always has the last word at sea, the officers, or the crew?" "The crew—'Aye, aye, sir!'"

Back(en) round. Of the wind, to change counterclockwise, a sign of bad weather on the coast. "It blowed fresh from the nothe-east all day, then backened round to the north'ard." See **Haul** (3), **Veer.**

Back and fill. A complicated procedure in navigation, involving the use of both wind and tide in a narrow channel. Ashore, the phrase means to vacillate. Its sea-derivation has been completely forgotten; a Midwesterner believed that it came from the operation of backing up and filling a dumpcart!

Back water. To row a boat backwards. In shore speech, it means to retract or hedge. To back out of an agreement may have the same derivation.

Backstays. See **Set up.**

Bail. From French baille, a bucket. To empty water out of a small boat. In coastal dialect, it means to fill or empty a utensil by scooping. "Quit bailing so much sugar into your coffee!" (Cf. bail out in aviation.)

Bait. A word common in landsmen's speech; for example, "You can't catch me with that bait"; "Grind your own bait" (be independent; or, in some connections, "Do your own dirty work"); "Fish or cut bait" (do one thing or the other; don't shilly-shally).

Ballast. Material placed in a ship to hold her upright when empty of cargo. Ashore, it has come to mean stability, good judgment, in people. The ballast of a railroad track is named for, and may in the first instance have consisted of, sand and gravel used to ballast ships.

Banian (Banyan) day. A day on which no meat was served on shipboard, usually Thursday; from Hindu merchant caste, Banians, who abstained from meat. It is occasionally used alongshore to describe short-commons or one of those days when everything goes wrong.

Banks, the. The Grand Banks of Newfoundland. A banker is a large fishing schooner using those waters.

Bar steward. Passengers' slang for bastard. Recent, and mostly British.

Barcarolle. A musical term, from the rowing songs of Italian boatmen. The most familiar barcarolle is that from Offenbach's *Contes d'Hoffmann.*

Bare poles, under. With all sails furled, while at sea, due to stress of weather. In coastal dialect it is facetiously used to mean naked, stripped.

Barge. A heavy, clumsy craft, with or without sails, used to transport bulky cargo. An older meaning was an officer's boat or ceremonial craft (for example, Cleopatra's barge). Ashore, it sometimes means a decorated float in a parade; while to barge is to move clumsily and awkwardly.

Bargee. (British). A man employed on a barge. Even sailors stood in awe of the quality of his profanity. Hence, to curse like a bargee.

Bark. Originally a general name for vessels of any rig, and still so used in poetry. Its sea use is now confined to a definite rig. The earlier spelling, barque, is still retained in England. The diminutive, barkey, is applied only to a well-liked vessel.

Barnacle. A marine growth on ships' bottoms. In coastal dialect, it implies signs of idleness or deterioration. "He's so lazy he's beginning to grow barnacles." See **Shellback.**

Barnacle Bill the Sailor, a song that raged on the radio a few years back, is a bowdlerized version of the old sea song *Bollocky Bill the Sailor*. (Ex *Abel* or *Abram Brown the Sailor.*)

Barney's brig, like. The complete phrase is "like Barney's brig, both main tacks over the fore-yard." It is used alongshore to express the extreme of disorder, something so bad as to be ridiculous.

Batten, Batten down. (French baton.) To cover securely against seas with fitted strips of wood or iron called battens; said of a hatch, etc. Alongshore, it means to protect with weatherboards. "I got the house all battened down for winter." (No connection with to batten upon, to grow fat, be a parasite.)

Battle the watch! A phrase of encouragement: "Don't give up; do the best you can." See **Watch.**

Beachcomber. A runaway sailor, originally from a whale ship in the South Seas. In shore speech, it means an unemployed person, an idler, especially a white man in a non-white country.

Beach, on the. Out of a ship, hence, unemployed; a phrase sometimes used in the same sense by landsmen.

Beacon. A sea mark to warn mariners from a rock or shoal; used ashore in various figurative ways.

Beaft. Same as **abaft,** and more common in later years, both at sea and alongshore. Perhaps a manufactured word, on the analogy of before.

Beam. The width of a vessel from side to side, or at right angles to her keel. Used in such expressions as forrard of the beam, abaft, or beaft, the beam. See **Abeam, Wind** (2).

In coastal dialect, broad in the beam is a facetious description of a person with heavy hips or buttocks.

Beam-ends, on the. Of a ship, careened at a sharp angle, so that she cannot right herself without cutting away the masts. In shore speech, the phrase means financially at the end of one's resources.

Bear. To lie in a given direction from the vessel. Used along-shore in the same sense. In shore speech, to bear upon, i.e., have a relation to, probably comes from this source; as also to bear to or toward, that is, to change course in the direction of. "After you strike the highway, keep bearing to the right." See **bear a hand.**

Bear down. To approach another vessel from the windward; a term originating in naval combat in the days of sail. In shore speech, to bear down upon one is to persist in argument; to be severe.

Bearing, Bearings. Position with relation to another known point or points, in navigation. In shore speech, the term is used to denote direction, influence, relationship; something has a bearing upon something else. To get, or lose, one's bearings means to become oriented or disoriented.

To bring a vessel down to her bearings is to **trim** her cargo so that she has no list and her keel is at the proper angle for easiest steering. See by the **head, stern.** In shore speech, to bring a person to his bearings is to bring him to reason, or put him in his place.

Bear up. To head the ship closer to the wind. John Milton was thinking of this when he wrote:

> I argue not
> 'Gainst Heaven's hand or will, nor bate a jot
> Of heart or hope, but still bear up and steer
> Right onward.

Perhaps from this secondary source, for there are unrecognized tags of Milton throughout the language, the phrase has come to mean in shore speech to be brave in sorrow, to show fortitude.

Beat. To make progress in the direction from which the wind is blowing by a series of zigzags. See **Tack, Windward.** In coastal dialect, a beat to wind'ard means a hard pull against adverse circumstances. See also beat the **Dutch.**

Becalmed. Held inactive for lack of wind. In shore speech, the term is used with a literary flavor for various states of inertia.

Beckets. Loops or handles made of rope. Used humorously in coastal dialect to mean the pockets of a man's suit. "He was strolling along with his hands in his beckets."

Beef, more. The joking call "More beef!" means that more hands are needed on a rope. It is occasionally heard on shore, when help of some sort is wanted.

Before the mast. Said of sailors in distinction from officers; the sailors' living quarters were in the forward part of sailing-vessels. The term may go back to the time of single-masted vessels. It is in common use alongshore.

Before the wind. See **Run** (2), **Scud, Wind** (2).

Belay. To make a rope fast by throwing turns around an upright pin, called a belaying-pin, which passes through a hole in the rail. Belaying-pins were weapons convenient to the hands of bucko mates; see **Capstan**-bar.

"Belay all!" or "Belay-oh!" is the order to stop pulling on ropes. "Belay your jaw!" means "Shut up!" See **Avast.**

Bellows, a fresh hand at the. A comment when a gale increases in violence is "There's a fresh hand at the bellows!"

Bells. See **Time.**

Below. Beneath the deck or in the deckhouses. "Go below the watch" is the conventional dismissal order at sea; "Below there!" is a hail, used similarly along the coast. "Go below and draw us a jug of cider."

Belowside. Down, down under. In less common use than **topside,** q.v. Pidgin English, brought home by sailors.

Bend. To affix, to place in position, said of sails and ropes. In coastal dialect, it is used similarly: "I got to bend them awnings today." It also means to put on a garment for the first time, as, "to bend a new dress." See also bend to one's **oars.**

Bends. Various knots, known alongshore as well as at sea by distinctive names, carrick bend, fishermen's bend, etc.

Beneaped. See **Tide.**

Berth. (1) The original meaning was "convenient sea-room, whence all later meanings are evolved; a good example of our love of nautical metaphor" (Weekley). From this comes the shore phrase, give it a wide berth.

(2) One of the subsidiary meanings is the space occupied by a ship swinging at anchor or in the dock; used also as a verb, one berths a vessel. This gives to land speech "he's got a good berth."

(3) A berth is also a built-in bedplace or bunk. The railroads

have borrowed the word in this sense to mean sleeping accommodations on a train.

Between-decks. A noun, meaning the upper section of a double-decked vessel's hold or cargo space. Alongshore, it is used jocularly to mean stomach. See **Hatch.**

Bight. A loop of rope; also, by extension, a cove or indentation in the coast line; used in both senses alongshore. Two ends and the bight means all there is of something, same as the landsman's two ends and the middle.

Bilboes. On old warships, sliding shackles on the deck in which men were chained up for punishment. In shore speech, in the bilboes was formerly used to mean imprisoned. (Literary, obsolete.)

Bilge. The rounded lower portion of a ship's hull. It is applied ashore to the bulge of a cask or barrel. See **Bung up and bilge free.** Of a ship, bilged means holed in the bottom by stranding; a seaman who fails to pass the examination for his next rating is said to be bilged.

Bilgewater. Stagnant water in a ship's bottom. In coastal dialect, any malodorous liquid; or, figuratively, distasteful talk, nonsense. The British have shortened the word to bilge.

Bill of health. See **Health.**

Bill of lading. See **Lading.**

Billow. A landsman's term for the seaman's **sea.** Entirely literary and poetic.

Bitter end, the. All authorities agree that this phrase relates to the end of the ship's cable attached to the windlass-bitts. When the anchor had been let out to the bitter end, there was nothing more to be done; if worse came, the cable would part and the ship drive ashore. In its shore use, to mean the last extremity, the sea origin has been completely forgotten.

Blackbirding. The African slave trade. In present-day speech, it means the fraudulent recruiting and exploitation of native labor in the South Seas and elsewhere.

Blanket. To cut off another vessel's wind by sailing close to windward of her, a maneuver in yacht races. Radio has taken the word over to mean interference with wireless signals. See also blanket **lee.**

Blockade, to run the. To carry cargo safely through a naval cordon. Hence, ashore, to make one's way skillfully through danger and opposition. (Cf. run the gantlet.)

Block and block (Sometimes abbreviated to Block and). Same as **Chockablock.**

B'longee. Belong; pidgin English brought home by sailors and chiefly used in the phrase, "No b'longee my," it isn't mine.

Blood in the scuppers. See **Scuppers.**

Blow. (1) A **gale**; usually modified by fresh, hard, or stiff.

(2) To knock down with the fists, as in the well-known sea shanty *Blow the Man Down.* The British colloquialism, blow me (if), and the American, I'll be blowed, come from this sea use of the term.

(3) At sea and alongshore, to blow when speaking of the wind means to blow heavily. By it came on to blow, a gale is meant. See **Breezen, Freshen** (1). To blow great guns means with a booming sound like cannonading, a sound often heard during cyclical storms. To blow like a man or blow right out endways also means with hurricane force.

Ashore, blow high or blow low equals let happen what may. A blow-hard signifies a loud-mouthed boaster.

Blow over. To clear (of the weather). In shore speech, it has gained the significance of being forgotten, losing importance with the passage of time.

Blow your horn. A standard exclamation alongshore, when someone blows his nose loudly, is "Blow your horn if you never sell a fish!"

Bluenose. A native of the Maritime Provinces of Canada, particularly applied to Nova Scotians.

Blue Peter. The code-signal P, a blue ground with a white square in the middle, hoisted at the fore to indicate sailing day. Of a person, flying the Blue Peter means that he is on the point of departure. British seamen call it the salt **horse** flag, in anticipation of that dainty.

Blue, till all's. A shore expression, to which some authorities give a sea-derivation, meaning till blue water is reached. This is probably an error. A list of tavern terms of 1650 lists, as a type of drinker "one that will drink till the ground looks blew."

Board. (1) (Anglo-Saxon bord.) The side or deck of a ship; two ships side by side are said to be board and board. The word is, however, used mostly in combinations, for example, **larboard, starboard,** for left and right-hand sides of a vessel; **freeboard, aboveboard,** for the portions of a vessel's sides above water; by extension, sternboard, for progress backward through the water; **overboard** for over the side and into the water. **Inboard** and **outboard** mean within and without the bulwarks. See these terms in the list for shore uses; see also **Aboard; In** (2).

(2) A board is also the distance made on a single **tack,** same as **leg** or **hitch.**

(3) To board a ship is to come over the rail and on to the deck. Ashore it is used commonly for the act of entering any means of transportation—bus, train or airplane.

Board in the smoke. To make an audacious move; to take by surprise. The term originates in old naval warfare or privateering.

Board, to go by the. Said of masts and spars broken off and swept over the side. In shore speech, it means to be completely ruined, a total loss. It often becomes garbled, however: "Everything I say here tonight falls by the board, unless— " said a political speaker recently.

Boat. In sailor's speech, this term is applied only to small craft, usually those propelled by oars, and never to large sailing vessels. A boat's length, in 'longshore parlance, is ten or fifteen feet. (Steamships used to be called steamboats, to emphasize the fact that they were not sail-propelled. This term is now generally reserved for river steamers.)

Small sailing-craft are customarily described by their rig, q.v., but sometimes by a combination with boat denoting the place where their design originated (Chebacco boat), their purpose (pilot boat), or, of steamers, their regular run (the Boston boat). Bradford states: "Boat, as distinguished from the general term ship, is constructed of bent frames, and a vessel or ship of sawn frames. (This is the opinion of a ship builder.)"

Sailboat applied generically to yachts, and boat alone as a name for a large sailing-vessel or steamship, are purely landsmen's terms; as is also the phrase, to miss the boat, meaning to get left, or fail to see the connection. All in the same boat signifies equally liable, or involved in the same situation. To burn one's boats is to take a course from which there can be no retreat. (Cf. burn one's bridges behind one.) See **Ship** (1), **Shipping, Sea boat.**

Boatswain (Pronounced bo'sun). A petty officer in the navy and **merchant marine.** He has given his name to the boatswain-bird, or tropic bird of the Southern Pacific, whose long tail resembles the boatswain's **marlinespike;** to the boatswain's chair, or swinging seat used in scraping and slushing down masts, now applied to the seat from which house painters do their work; and to the boatswain's locker, a place for storing odds and ends, frequently so designated in houses alongshore.

Bold-to. Of a coast, shelving steeply under water; safe to approach.

Booby hatch. A **hatch**way leading to storage space under the poop deck. The term has recently made its appearance in shore slang, probably via the Navy, to mean a hospital for the insane.

Boom (Dutch). A spar used principally to extend the foot of a sail, in distinction from a yard, from which the head of a sail is suspended. Says Professor Chase: "Boom is a mighty interesting word. Originally it meant a tree (German baum). The Anglo-Saxon is beam, a squared log; the Dutch is boom, a rounded spar. Is that an indication that we owe something of the **fore-and-aft** rig to the Dutch?"

The lumberman's boom of logs is probably from the nautical boom. With **studding-sails** (q.v.) boomed out, a ship was said to be "booming along." From this perhaps comes the shore boom, a period of prosperity, when things are booming, (though some believe that this shore term comes from the boom of an explosion).

Booze. Alcoholic liquor; or to drink to a state of drunkenness. From an obsolete sea term, bowse (or bouse) the can, which meant to drink heartily, or to haul a rope taut with block and tackle.

Bore. A periodic tidal wave in a river; an eagre. The one most dreaded by seamen occurs on the Hoogly River in India.

Bose. See **Forms of address.**

Bottom. A vessel considered as a receptacle for cargo. Shakespeare says: "My ventures are not in one bottom trusted." See **Bilge, Sail** (2).

Bound to, Bound for. (Old English boun, ready, prepared.) Terms denoting a ship's destination. Alongshore, they are used generally. "Bound for the village, are you?" See **Outward bound.**

Bowers. The ship's heaviest anchors, the larger being her "best bower." This term, occasionally used figuratively by landsmen to mean the same as trump card, may come from this nautical source, but more probably from the game of euchre, in which the two bowers are named for the German bauer, peasant, and not for the ship's bow-ers.

Bowl along. To sail fast and merrily. It may come from, or be the origin of, the game of bowls ashore.

Bowline (Pronounced bo[ō]-lin). (1) An extremely useful sailor's knot that will not slip. It has many variations.

(2) In ancient days, the fore-sheet; later, a light rope used to steady the leach of a sail.

On a bowline means sailing closely by-the-wind.

Bows-on. Said of another vessel headed directly for the speaker's. Alongshore, it means headfirst. "He run into me bows-on."

Bows-under. Said of a vessel plunging head-on into heavy seas. Alongshore, it signifies making difficult progress, as a sleigh in deep snow; or figuratively, being overwhelmed with work to do.

Boxhaul (around). This originally meant tacking ship with **sternway** on instead of **headway,** a dangerous emergency maneuver. Later, it has been carelessly used to mean, of the wind, to whiffle about; of the yards, to be continually bracing them to meet slight changes of wind, as in the old song:

Oh, now we are sailing up on to the Line;
When I think on it now, sure, we had a good time,
Those sea-boys box-hauling the yards all around
For to beat that flash packet called the *Thatcher McGown!*

Alongshore, the term means to be changeable and capricious.

Boy to do a man's work. The complete phrase, at sea and alongshore, is "Don't send a boy **aloft** to do a man's work." Certain light duties, such as furling the royals, were reserved for the ship's boys; the men were supposed to handle the heavier canvas. The phrase is used sardonically alongshore when some particularly feeble and unsuccessful effort has been made. It is occasionally used by landspeople in the same sense, but omitting the word aloft.

Brace. (1) The rope used to swing the yards on a square-rigged vessel. The shore expressions, take a brace, and get a brace on, probably come from this source, though brace itself is not a word of sea origin.

(2) To brace up is to cant the yards so as to get the full advantage of the wind; superlative, brace up sharp. Ashore, the term is used, with no recollection of its origin, to mean summon courage. See **Bear up, Set up, Splice** the main brace.

Brace of shakes. This is given in one list of British nautical terms as meaning the time it would take a sail to flap twice. This is a doubtful derivation. Cf. two shakes of a lamb's tail. The American prefers "a couple of shakes."

Break. In the sailor's vocabulary, a rope does not break, it **parts.** Only the sea breaks, but see break **adrift,** break **bulk.**

Break away. Of the weather, to clear after a storm. "It's going to break away before noon."

Breaker (From Spanish barrica). A water keg in a ship's boat. Newspaper accounts of sea disasters sometimes have it beaker—"a perversion due to ignorance," according to Weekley.

Breakers. Heavy waves rolling up on a reef or shore. The word appears in several shore phrases: "Breakers ahead," or "Look out for breakers!" warning of danger; and "in the breakers," on the verge of disaster.

Break out. (1) To loosen the anchor from the bottom preparatory to weighing it.

(2) To bring up from the hold, as cargo, to unpack, as stores. Alongshore, it means to bring newly into service. "Guess I'll break out those new dishes for the party." See **Bend.**

Break the ice. A shore expression, said to come from Arctic exploration.

Breeze(n). (1) The sailor's breeze signifies more wind than the landsman's "every passing breeze," which to the seaman, would be merely light **airs.** A stiff, fresh, spanking or whole-sail breeze means good sailing weather; but a breeze o' wind signifies a moderate **gale.** The breeze was straight up and down, however, denotes a dead calm. See also **White-ash-breeze.**
(2) To breeze, or breezen, is to **blow** harder; to increase nearly to the velocity of a gale. Variations are breeze up fresh, breezen on.

Bricklayer's clerk. One of many terms for an inefficient seaman, especially one who is fastidious or seems to feel too good for the job. (British seaman's term.)

Brig, the. The prison on a warship. In the brig, via Naval slang, signifies in jail.

Bright. On shipboard and alongshore, polished or varnished. Brightwork means polished copper or brass fittings; brightwood is cabinet woodwork or carving, varnished or shellacked, not painted. See also **Lights, Lookout.**

Bring aft. To promote from the fo'c'sle to an officer's berth. Used in the same sense alongshore.

Bring home. See **Home.**

Bring up. Of a ship, to come to a sudden stop, due to grounding. To bring up all a-standing is to go on shore with all sail set. To bring up with a round turn is to bring to a sudden and effective halt, as by snubbing a rope around the bitts or pin. See **Snub.** Alongshore, to bring up short or all a-standing means to surprise or disconcert; while to bring up with a round turn means to rebuke sharply. In all these phrases, fetch can be substituted for bring. "The parson fetched him up with a round turn and no mistake!"

Broach-to. Of a vessel, to get off her course when running before a heavy wind, so that she falls into the trough of the sea, and is in danger of capsizing. Alongshore, it might be said of a vehicle that gets in the ditch. See **Capsize.**

Broadside. The naval maneuver of firing off at once all the guns on one side of a ship. In shore speech, it denotes a vigorous and effective argument or attack, especially in the newspapers.

Broken up, to be. The final destiny of a ship. In shore speech, it means to be deeply moved or grieved, frequently with "all."

Buccaneer. A freebooter or pirate of the Spanish Main, 17th-18th centuries. The modern figurative use, to mean a lawless person, is chiefly literary.

Bucket. No pail was ever used on shipboard, everything of that nature is a bucket. See **Draw-bucket.** In coastal dialect, the same preference is shown; water-bucket, tar-bucket, bucket o' clams. Pails are utensils of farmers; milking-pail, lard-pail, swill-pail, etc.

In shore speech, pail is preferred, and bucket, when used, means a wooden container. (To kick the bucket, meaning to die, has been suspected to be of sea origin, but this has never been proved.)

Bucko. Designates an officer who used brutal methods of discipline, a bucko mate. Alongshore, it means a bully, especially in the sardonic appellation "My bucko!"

Bugger. To people who speak by the dictionary, this is a highly obscene word, but as used by sailors, it carries no shade of its actual meaning. It is seldom used as a term of address, but rather of reference, in telling stories; and it carries about the meaning of fellow or rascal.

To bugger is to confuse or perplex; "I'll be buggered!" an exclamation of mild astonishment. That seamen—at least, those of fifty years back—had not the remotest idea of the real meaning of the word is amply proved (as anyone who knows sailors will agree) by the fact that they used it freely in the presence of respectable women. (Cf. the English word beggar, which is evidently an attempt to soften the term.)

Build. Model, figure. "She's got a good build on her." This use of the word is suspected to be of sea origin.

Bulk. Originally meant cargo in a ship; still preserved in this sense in the phrase, to break bulk, i.e., begin unloading. Shore uses to mean size, mass, are taken from the earlier sea use.

Bulkhead. A perpendicular division in a ship's hold, separating cargo stowed in bulk. In shore speech it means a partition, retaining wall.

Bully. First-rate. The earliest use of this term was on land, but it soon became engrafted in seamen's speech. It has become common and popular in shore slang.

Bully beef. A sailor's word for canned meat, adopted by the English and American armies in the first World War. It is sometimes stated to come from the French bouilli, boiled beef; but is much more probably from bull-beef, on account of its stringiness.

Bulwarks. The sides of a ship above deck level. In shore speech, the word signifies, figuratively, a strong defense. "The surest bulwark against anti-Republican tendencies," thought Thomas Jefferson, was states' rights.

Bumboat. A small boat which peddles supplies to ships in harbor; familiar to landsfolk through the Gilbert and Sullivan opera, *Pinafore*.

Bumpkin. Boom-kin, a short boom. A timber projecting from the hull to which tackles may be attached. How, or if, it is related to the country bumpkin is not known.

Bung up and bilge free. Properly disposed or packed. From the whaleman's description of the proper stowage of casks of whale oil.

Bunk. A built-in bedplace on board ship. See **Berth** (3). In shore speech, a bed; also, by extension, to sleep, bunk in with. One etymologist says that to bunk, or do a bunk, meaning to decamp, may have come from this source, and have meant originally to abscond by sea.

Bunting. The woolen material from which ship's flags are made; hence, her flags in total. The material is also sold by drapers ashore, often under its French name, étamine. The name "bunting (bolting) cloth" was probably originally of shore origin.

Buoy. A sea mark; or a float for lifting objects in the water. Hence, in shore speech: buoyed up, by hope and the like; and buoyant, hopeful, optimistic.

Burdensome. Of a ship, having a good carrying capacity; used alongshore in the same sense. (The shore use to mean onerous has no connection with the sea.)

Burgoo. The sailor's name for oatmeal porridge. See Thick.

Buying into (a ship). See **Vessel property.**

By and large. Sailing by and large is with a fair or leading wind. See **Large, Wind** (2). Ashore, although its sea origin has been forgotten, the expression signifies "in general, in most respects."

By the wind. As close to the wind as it is possible to sail and still keep the sails full. See **Wind** (2).

Cabin. Living accommodations aft for officers and passengers. On modern steamships, cabin is equivalent to **stateroom.** Ashore, it sometimes means a compartment on a sleeping-car.

Cabin window, through the. A shipmaster who began his sea career as an officer is said to have "come in through the cabin window." See through the **hawsepipe.**

Cable. A heavy rope; or the anchor chain (which used to be made of rope; the name has been carried over into the modern chain-cable). In shore speech, cable has been adopted as the name for the submarine telegraph; by extension, to cable is to send a message over it. To cut one's cable is to decamp without ceremony; to slip one's cable is to die.

Cable's length. A sea measurement variously given as 100 to 120 fathoms. See **Fathom** (1). As a unit of sea distance, however, it is approximately one-tenth of a sea **mile,** 100 fathoms or 600 feet. The word was used in early surveying at sea; it also appears in early literature as applied to land distances, though not in precise connections, "scarce a cable's length," etc.

Caboose (Dutch). The cabin or deckhouse on a small vessel; the **galley.** Ashore the word may mean a shack, or the last car of a freight-train, fitted for the accommodation of the crew. (Caboose as a variant of calaboose, meaning jail, is probably from the Spanish calabozo, and not of nautical origin.)

Call. Used on shipboard only to mean summon from below decks, as in calling the watch to come on deck for duty. The captain orders the officer of the watch to call him if the weather changes, for example. But people on deck either **hail** or **sing out** to each other.

Call at. Same as **touch at,** q.v.

Cally-o. Pronunciation alongshore of Callao, Peru. All Cally-o is a phrase meaning all right, or even better. Callao is a port full of delights for seamen.

Calm. Windlessness. Variants are a flat calm, a dead calm.

See **Becalmed, White-ash breeze.** Used figuratively in shore speech—the calm before the storm, etc. Other names for a calm are Irish hurricane or Paddy's hurricane.

Camel. A sort of sledge for heaving boats or small craft over a shoal or bar. In coastal communities, it is said of a light-weight, "I guess he'd get over the bar without camels."

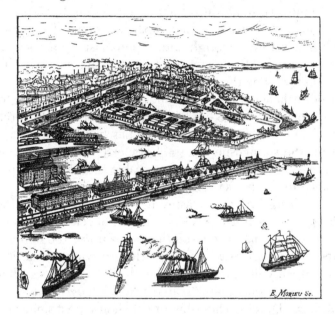

E. MORIEU Sc.

Can do. Signifies agreement, assent. The negative is "no can do." Pidgin English, brought home to the coast by sailors.

Canvas. French canevas, sailcloth (before that going clear back to the Greek kannabis, hemp). The fabric from which a ship's **sails** are made. The sails of most European vessels are made of hemp; while those of American vessels are of cotton. See also **Duck, Reef** (2).

By extension, a ship's set of sails is her canvas.

> As ships becalmed at eve that lay,
> With canvas drooping—

Cape, the. Cape of Good Hope or Cape Cod, according to the context—never Cape Horn. See The **Horn.**

Cape Flyaway. A cloud-bank on the horizon having the appearance of land.

Cape Horn, around. In whaling communities, this has the special significance of being away on a whaling voyage.

"I'll tell your father, boys," I cried,
To lads at play upon my lawn.
They chorused back, "You'll have to go
Around Cape Horn!"

A Nantucket housewife whose husband had hung around the house too long to suit her, is reputed to have said to him, "One or t'other of us has got to go around Cape Horn pretty soon, John, and it ain't a-going to be me." See The **Horn.**

Cape Horn fever. Malingering; cowardly avoidance of danger.

Cape Horn snorter. A heavy gale of the sort experienced off Cape Horn; usually with "a reg'lar old."

Capful o' wind. See **Wind** (1).

Capsize. The derivation of this word is unknown. It has not been found before 1760; up to that time the nautical term was **overset.** Said of a ship, it means to turn completely over and float bottom up.

Alongshore, it is used about upsetting anything. "Look out and don't capsize that pan o' milk!" "His load of hay capsized."

Capstan (Pronounced caps'n; formerly spelled cabestan). An upright mechanical device used to operate the windlass in weighing anchor and for other heavy jobs; also, the heavy timbers edging a wharf.

The wooden capstan-bars used to drive the capstan round are ugly weapons in a melee; hence, caps'n-bar hash and belaying-pin soup, meaning physical abuse.

Captain. Southern French capitaine; it may have been used first as a naval rather than a military title. See **Forms of address.**

Captain's walk. A gallery around a cupola (pronounced cupalow) on mansions of the type originally built by retired shipmasters.

Careen. From the French caréner. (1) To **heel** over or **list.** (2) To heave a ship down on a sheltered beach in order to remove weeds, etc., from her bottom. A place where this was frequently done gained the name of Careenage, sometimes given a French twist and abbreviated, as in Bridgetown, Barbadoes, to "The Ca'nash." See **Heave down.**

Cargo. A borrowed Spanish-Portuguese word, meaning load. The earliest use of this word related to loading carts (from Latin carrus), but in modern English, it is reserved for goods transported in a vessel, and its figurative uses in shore speech, a cargo of good wishes, for example, have this derivation.

Alongshore, a person who has a cargo aboard is intoxicated; a full cargo signifies that there is no more room in a conveyance.

Carry away. To break suddenly; said of a rope or spar. Used alongshore in the same sense: "The clothesline carried away when the wind rose." See **Part.**

Carry on, or **Carry sail.** To keep a little more sail set than the wind justifies, so that the ship dances and plunges about. Carry on or carry under is the motto ascribed to a **driver.** See **Crowd on, Crack.**

In shore speech, to carry (all) sail means to live high, be in full career, or to put on airs; while to carry on, or carry on regardless, means to joke and romp uproariously. Sometimes the participial noun is used, "Such carryings-on!"

Cast about. To try different courses when in doubt as to the ship's position. In shore speech, with its sea origin forgotten, it means to grope, to try to find a way out of a dilemma.

Castaway. A shipwrecked person; an outcast. In its earliest shore use, the word did not refer to the sea. See **Derelict.**

Cast off. Of a rope or line, to untie; to loose from pin or bitts; common in the same sense alongshore. It may be related to the shore term, meaning to discard, abandon. See **Adrift, Away, Overboard.**

Cat-o'-nine-tails. The instrument used for flogging in the old Navy. Frequently abbreviated to cat. A playful threat to 'longshore children is "I'll take the cat-o'-nine-tails to ye!"

A phrase common in shore speech is: "not room enough to swing a cat." Shore users, under the mistaken impression that this refers to poor pussy, have added "by the tail."

Catspaw. A localized light breeze on a calm day, shown by a streak on the water. In the British Navy, a recruiting officer was formerly called by this name.

Catwalk. An elevated bridge connecting the midship house with the forward or after part of the vessel. The term is used ashore to designate similar structures in industrial plants.

Caulk off. Alongshore, to caulk off is to take a nap, probably likening the sound of snoring to that made by caulkers' mallets.

Chalk, not by a long. This phrase, which has been erroneously ascribed to the sea because of its resemblance to "not by a long **sight**," probably comes from the practice of chalking up customers' accounts on the walls of drinking-places; or possibly from scoring points in games.

Changee for changee. The other way around, or "let's exchange." The complete phrase is "changee for changee, black dog for white monkey." Pidgin English, brought home from China by sailors.

Channel, the. The English Channel. The "chops of the Channel" is the western entrance, where Land's End and Ushant oppose each other like jaws.

Channel. French, from the same Latin source as **canal**, q.v. It was not originally restricted to navigable waters, but meant waterways in general, the beds of streams, etc. The word has come to mean principally the navigable reaches in rivers and harbors, or some constricted passage through which vessels may safely go. In shore speech, it is used to describe any means of access; for example, regular banking channels.

Chantey. See **Shanty.**

Charley Noble. The galley stovepipe. Sometimes used for a stovepipe (not chimney) alongshore.

Chart. (1) A map used in navigation, showing soundings and compass-readings. It has been adopted ashore as the name of certain forms of statistical graphs.

(2) To map, lay out. This dates back to the time when coasts were surveyed by explorers and traders as they went along. To chart a course (of action) is now completely a landsman's expression. See **Uncharted.**

Charter. As a noun, charter is not of sea origin, but to charter means of transportation related from the beginning of its use only to ships; so that modern use ashore to mean hiring or engaging a special train, car, bus or airplane comes from the sea.

The charter party, or agreement between shipper and carrier, comes from the French nautical term charte partie, a document of which one-half was retained by each of the parties to the transaction.

Cheerly. Quickly, with a will; an obsolete direction to sailors at work. A nonsense-burden recurring in French sea songs, "Oh! Célimène," has been shown to be the English "Oh, cheerly, men!" It is sometimes erroneously written "cheerily," as in the shore song *Nancy Lee*, with its chorus, "Cheerily, my lads, yo-ho!"

Chew the fat. According to Bowen, this is a sailor's term, meaning about the same as to **growl,** q.v., and referring to gristly salt beef which requires much chewing before it can be swallowed.

In shore slang, it means, rather, to talk aimlessly and for a long time.

Chief. See **Forms of address.**

Chinse. To caulk or fill in a very narrow seam; probably related to chink.

Chips. See **Forms of address.**

Chockablock. Said of a **purchase** or **tackle** when both blocks have been brought together and no more slack can be taken on the rope. "I'm chockablock and belayed," said an innkeeper on the Maine coast, in turning away a would-be guest. Ashore (colloquial), it means completely; often abbreviated to chock, as in chock-full.

Chock off. To fill in around an object in stowing, in order to keep it from moving. Used in the same sense alongshore.

Chop. See **First-chop.**

Chop-chop. Quickly, in a hurry. Pidgin English, brought home by sailors.

Chops of the Channel. See The **Channel.**

Chopsticks. Chinese eating-implements (k'wai-tsze, the nimble ones). Translated into pidgin English, chop (quick) and sticks; brought to this country by sailors. Weekley says the term was used by Dampier in the 17th century.

Chow. Food; or to eat. Pidgin English, brought home by sailors. Chowchow is a kind of chopped pickle made in New England.

Chowder (French chaudière, a cauldron). A stew of fish or clams with onions and potatoes, popular at sea and ashore. The name was brought to the Maritime Provinces of Canada by Breton fishermen, who were the earliest settlers. Chowder is made with milk in Maine and Massachusetts, with water and added tomatoes in Rhode Island and further south. A Maine solon recently introduced a bill into the state legislature making it illegal to add tomatoes to chowder within the State of Maine, the penalty being that the offender must dig a barrel of clams at high water!

Clam, happy as a. Alongshore, this phrase is happy as a clam at high water; that is, when no enemy can reach him. The landsman has abbreviated it to the meaningless happy as a clam.

Clap under hatches. See **Hatch.**

Claw off. To beat off a lee shore; hence, to get out of a threatening situation.

Clean bill of health. See **Health.**

Clean sweep, to make a. Of a sea breaking over the rail, to wash overboard all movable objects on deck. The phrase is common in shore speech, meaning to get rid of everything; to start fresh.

Clear. (1) Of the weather, affording good visibility. Of ropes, not entangled, free-running. The opposite of **foul,** in both these connections. The shore phrase "in the clear," that is, free from blame, probably comes from this sea use. See the **coast is clear.**

(2) To clear the ship is to complete the formalities necessary before leaving port. She is then said to have cleared, or been cleared, for her port of destination. This is the origin of the injunction, "Clear out!" Alongshore, a child would be told to "clear for home."

(3) In passing an obstacle or another vessel closely, the ship clears it; in putting to sea, she clears the land. The old-time naval vessel in preparing for battle, cleared for action or cleared the **decks**—phrases common in shore speech. See also **Steer** clear.

Clearance papers. Official permission to sail. Alongshore, to take out clearance papers for a place is a facetious way of announcing one's destination; while to get one's clearance papers is to be dismissed. (Clearance meaning space or head-room is probably not of nautical origin.) See **Papers.**

Cleat. A small wooden contrivance upon which to belay a light line. Used in the same sense alongshore.

Clew to earing, from. Names of the opposite corners of a square sail. Alongshore, it signifies completely, all over, from one end to the other. See **Alow** and aloft, from **stem to stern-post; from truck to keelson.**

Clew up. To haul up the corners of sails preparatory to furl-ing. Alongshore, to lift and fasten (sometimes said facetiously of a woman's skirts). See **Trice.**

Clinker-built. Of a boat, having overlapping strakes; a term of general approval alongshore.

Clipper. A type of vessel in whose lines carrying capacity was sacrificed to extreme speed. Alongshore, a rollicking person is called a clipper, or for short, a clip. Clipper-built, of a woman's figure, means neat and trim. See **Frigate.**

Close aboard. See **Aboard.**

Close-hauled. Same as **by the wind.** See **Wind** (2).

Close quarters. In a naval engagement, this means in hand-to-hand combat. See **Board** and board. Ashore, it is applied to restricted space.

Close shave. A near-disaster. Probably of nautical origin.

Close with, to. To engage another ship in battle. Ashore, it signifies to strike a bargain.

"Coals to Newcastle." Newcastle is a North Sea port the chief industry of which is mining and exporting coal. Hence, to ship a cargo of coal there, "carry coals to Newcastle," would be highly superfluous. The phrase is firmly embedded in shore speech.

Coast. To make short trips from port to port. Alongshore, to coast around means to wander aimlessly.

Coast, the. The Pacific coast of North America. But to come on the coast means to reach the vicinity of the Atlantic coast ports. See **West Coast.**

Coaster. A small vessel trading from port to port along the coast, a shipmaster who never went **deepwater.** See **Appletree-er.** A small tray to set under a drinking-glass is called a coaster, which may or may not be connected. See **Schooner.**

Coast is clear, the. A shore expression, meaning that one may proceed without fear of interruption; probably referred originally to a blockading squadron.

Cockbill. To brace yards unevenly or in opposite directions. Alongshore, to hang askew. "I don't like to see pictures all cock-billed on a wall."

Cockroach (Spanish cucaracha). Sailors first made the acquaintance of this life-long friend of theirs in the early exploration of the West Indies. Captain John Smith writes of "a certain Indian bugge called by the Spaniards a cacaroatch." There can be little doubt that the water bug of our coastal cities, a small cockroach, was brought to the ports in ships.

Coir. See **Kyar.**

Collision, to be in. Ships do not collide; they are "in collision with" each other. See **Run down.** This usage is followed alongshore, in regard to vehicles, persons, etc.

The Rules of the Road to prevent collision are learned by heart by every young seaman in sail. Those for daytime use are:

If close-hauled on the starboard tack,
No other ship may cross your track.
If on the port tack you appear,
Ships running free must all keep clear;
While you must yield, when running free,
To ships close-hauled upon your lee.
And if you have the wind right aft,
Keep clear of every sailing craft.

(For rules at night, see **Lights.**)

Colors. Although the plural form is used, this word refers to a single flag, the ensign, or flag of the country from which the ship hails. It is hoisted or set, dipped in salute, or hauled down, as occasion demands, by means of a light line called the signal-halliards. No other flag may be flown above it except the chapel flag, to show that a church service is being held on board, but code signals showing the ship's name may be hoisted under it. See **Signals.**

The colors are set upside down as a signal of distress. See **Union down.** When set part-way up the halliards, "at half-mast," it means that there has been a death on board. Shore use of this term and practice is probably derived from the sea.

Numerous phrases current in shore speech hark back to the use of the ship's colors in time of war, to identify or conceal her nationality, and to show surrender or refusal to surrender. Examples are:

> To come off with flying colors;
> To haul down (or strike) one's colors;
> To go down with colors flying (or flying colors);
> To nail one's colors to the mast;
> To set (or hoist) one's colors;
> To show one's true colors;
> To sail under false colors ("put a false color on" something).

There are also "under color of" and "stick to one's colors," but these two may be military rather than naval in origin.

Come day, go day, God send Sunday. Sunday at sea is the sailor's one day of leisure, eagerly anticipated. This phrase is applied alongshore to people who work in a listless and uninterested manner, the implication being that they are clock-watchers.

Come down upon. An old term of the sailing Navy, meaning to attack from the windward. See **Weather** gage. In shore speech, it means to censure, rebuke, give one a blowing-up. "The boss came down on me hot and heavy."

Come up. To slacken up a rope that has been hauled too taut. To come up the **capstan** is the same as to surge it. The Elizabethan phrase "Marry, come up!" meaning stop, hold, probably came from this source. See also come up **thick.**

Commission, in or **out of.** In or out of active service (said of naval vessels). In shore speech, it means active and in use, or the reverse. To put a person out of commission means to render him ill or helpless. See Roving commission.

Company. Ships sailing along the same course and near together are said to be in company; when their courses diverge, they part company.

In shore speech, to keep company, as sweethearts; also to fall in company with (sometimes shortened to fall in with), or meet someone accidentally, may have this origin. To part company means to be divorced, break up a business partnership, or, facetiously, to come apart. "His shirt and pants parted company."

See **Ship's company.** (It is entirely possible, of course, that all these phrases were first used on the land.)

Compass. A magnet attached to a swinging card bearing a circle marked off into 32 points. The compass was originally called in English the needle or the stone (see **Lodestone**), and seems to have received its present name not earlier than 1500—probably through some analogy with compasses, mathematical instruments used to draw a circle.

The compass card floats in a square box; hence, to box the compass is to name all the points, following the edge of the box. See **Points of the compass.** In shore speech, to box the compass means to go completely round a subject; cover it thoroughly (for an illustration, see page xi). From all points of the compass means from every direction; while to speak by the card—that is, advisedly—probably refers to the compass card.

(The phrases to fetch a compass round, or to encompass one's ends, for example—even Tennyson's "compassed by the inviolate sea"—probably come from the mathematical instrument mentioned above, and not from the mariner's compass.)

Comprador. A business agent. Portuguese, via pidgin English, brought home by sailors. (Dubash, an, East Indian word of similar meaning, is understood but not used along the coast.)

Con. To guide the helmsman, usually by shouted directions from aloft. The word is used alongshore in a similar sense: "I don't know the way, but I'll drive you up there if you'll con me."

Consign. To forward cargo to a given consignee in care of the vessel's owners. The land meaning of "to entrust" is the earlier; but just as the railroads at their beginnings took over wholesale words relating to transport by sea, it may fairly be claimed that consign, as well as consignment, consignor and consignee, entered the vocabulary of modern land commerce from water-borne trade.

Contraband. Goods subject to seizure in time of war. The word contraband in English dates from the illicit trade with Spanish possessions in South America in the 17th and 18th centuries, and is therefore of sea origin.

Cook. Many coastal proverbs make the ship's cook the butt of rough jests: "God made the vittles but the devil made the cook." "We must take it as it comes from the cook" is used to express humorous resignation about everything else but food. See **Son of a sea cook, Galley** news.

Cootie. A body louse; a sailor's name for the insect, which was taken over by the armies and given currency in civilian speech during the first World War. It comes from the Polynesian word kutu, meaning a parasitic insect.

Copper-bottomed, Copper-fastened. Copper was preferred for metalling wooden ships because of its superior resistance to corrosion, marine growths, etc. Hence, at sea and on the coast, the term means well-made, handsomely finished. A returned sailor, describing the roofs of the houses in Beirut, said they were "copper-bottomed with sheet-lead."

Correct the Declination. See **Declination.**

Counters. The overhang of a ship's **quarter.** Alongshore, hips are facetiously called counters.

Couple of shakes. See **Brace of shakes.**

Course. The direction in which it is desired to make the ship sail. The master lays, sets, or shapes the course for the ship to steer. She then holds to, or is on, her course, or may be off her course if the helmsman is unskillful.

Many figurative phrases come from this nautical use of the word, the meanings of which are obvious. Examples are: to lay, set or shape a course; to steer, keep to, hold to, a (safe) course; to be off, or on, the right course ("keep thy course aright"); to alter, change, shift one's course; to run its course (of a disease, an adverse condition). See **Steer, Run.**

Crack. A term of approval of ship or crew. It was probably originally a land word; but to "crack on sail" is purely of the sea. See **Carry on, Crowd on.**

Craft. Any type of vessel, large or small. Alongshore, it may also mean a woman. "She's a likely-lookin' craft." In Trinidad, "craf" means sweetheart, girl friend.

Crank, Cranky. Of a ship, defective in design so that she has a tendency to **heel** over even when sufficiently ballasted and not overdriven. A crank vessel is a mean beast to handle. In this sense, the word comes directly from comparatively late Dutch-Scandinavian words of the same meaning.

The ultimate derivation of the word is, however, extremely involved and obscure. The Century Dictionary lists six different groups of uses of crank, all of which are interrelated in origin. But none of the six covers the way in which the word is used in coastal communities, to mean either irritable and fractious or definitely ugly; that is, cross-grained and quarrelsome. This seems clearly to come from the crank ship.

As used in shore speech to mean a person who is ill-balanced and flighty, or one who is a monomaniac in pushing some impractical purpose or idea, the sea-derivation would be harder to establish. This shore crank probably goes back to two Anglo-Saxon and Middle English words, one meaning a crooked tool used by weavers, the other, weak and infirm. The modern shore sense may have been influenced also by the later nautical meaning.

Crew. Usually means the foremast hands on a vessel, but under some circumstances it legally includes both sailors and officers. All shore uses of the word come from the sea. As applied to work crew, train crew, wrecking-crew, etc., the term is literally employed; in other connections, it has a disparaging flavor, a motley crew, and so on.

Crimp. The keeper of a sailors' boarding-house, who formerly supplied crews, and who fleeced the sailors of all that they had earned. See **Landshark, Shanghai.** The word is said to have come from an old cheating-game. It is occasionally confused by landspeople with pimp, which is no insult to either of these gentry!

Cross-currents. See **Current.**

Crossjack-eyed (Pronounced crojick). Squinting; but for no known reason. The crossjack is a perfectly symmetrical square sail on the mizzenmast.

Crowd on. The same as **carry** on, q.v. In shore speech, to crowd on sail is understood to mean making unusual or increased efforts.

Cruise (Dutch). To voyage without a stated destination, as whalers, explorers, privateers. (But a cruiser is a naval vessel.) Modern steamships cruise when taking passengers on a pleasure voyage. In shore speech, to cruise is to wander about, or to explore. A timber cruiser is an expert woodsman who estimates stumpage in advance.

Cumshaw. Chinese; a beggar's term meaning alms. In pidgin English it means, however, an honorable gift; something thrown in on a trade. The term was brought home by sailors.

Current. A moving stream of water within the sea, not periodic as the tide. Currents are not dependable, and may set the mariner out of his reckoning.

The term appears in many figurative phrases ashore: the current of one's thoughts, of events, of the times, and so on. One makes difficult progress against the current or goes easily with the current. Conflicting impulses or purposes are cross-currents.

Curry (Tamil). A favorite dish in coastal communities, and now well known to shore people as well. Curry powder was originally brought home from India by sailors. In Trinidad, it is called talcoree, and is believed to have aphrodisiac properties.

Cut adrift, away. See **Adrift, Away.**

Cut and run. Reference is to cutting the anchor cable. It is now chiefly a shore expression, meaning to make a hasty getaway. See **Cable, Stick.**

Cut loose from. A shore term which may have a sea origin.

Cut out. A naval term for attacking a ship at anchor by means of small boats. This is probably the origin of the shore colloquialism "to cut out" another man with his girl.

Cutwater. The extreme forward plank of a ship's bow. Alongshore, it is sometimes jokingly used to mean nose. In engineering, the upstream edge of a bridge pier is the cutwater. (See page xvi.)

Cyclone. The generic name for cyclical storms, but by seamen reserved for those met in the Indian Ocean.

Dago (From Diego, a Spanish first name). A person belonging to any of the Latin races, except Frenchmen; at sea they are "Frenchies." (The disparaging term "Frogs" is not of sea origin.)

Dandyfunk. A sailor's term for a pudding made of cracker crumbs, **slush** and molasses.

Davy Jones' Locker. The bottom of the sea, where everything goes that is thrown overboard, including the bodies of dead sailors. Many sea superstitions involve Davy Jones, the goblin of the deep, but only his locker appears in the speech of landspeople. See **Fiddler's Green.**

Dead. An intensifying word, used to denote exact bearings: dead ahead; dead astern or aft; dead to windward or leeward. This is a nautical adaptation of a common shore use of the word, as in dead right, dead wrong, etc. See **Due, Reckoning.**

Dead horse. On British ships, the dead horse is an effigy thrown overboard when one month out, with suitable ceremonies. It celebrates the working off of one month's advance wages received by the crew when signing the articles, and usually promptly confiscated by the boardinghouse keeper (see **Crimp, Landshark**) for alleged back debts. In shore speech, it means the paying off of any sort of debt. "We've at last got rid of the dead horse on the club-house."

Dead whale or a stove boat, a. The whaleman's slogan. Alongshore, it means "Do or die," or "Go ahead whatever the consequences."

Deal beach, rolled on. Pock-marked; from the sharp gravel on Deal Beach, England. (British term.)

Deck, Decks (Dutch). The horizontal partitions separating the sections of a ship's hold, and particularly the top planking. The Anglo-Saxons, using open ships, had no need for the word; the fact that it comes into English as a borrowing from the Dutch may indicate that they were the first to use decked vessels.

The word appears in many shore phrases. Always on deck means alert, not to be taken by surprise; but right on deck has, at least in coastal speech, a slightly sarcastic flavor, to mean pushing, too much in evidence. The Navy of today contributes "Hit the deck!"—a call to get out of bed.

In baseball, the man next in batting order is on deck. In certain suburbs we find three-decker houses reached by two-decker busses; a large ice-cream cone or a multiple-layer sandwich is a double-decker. Some hotel or penthouse roofs are called decks. See **Clear** the decks, **Quarter-deck.**

Deck load. Cargo carried above-deck. Alongshore, it is often used to mean too much to drink. "Lafe had quite a deck load aboard last night."

Deck seam, to walk a. To demonstrate that one is sober; same as the shore "walk a chalk line."

Deckspike. A large nail used to fasten deck planks. See **Spike.** The term is used alongshore in the self-explanatory phrase, "He'd snap the head off a five-inch deckspike."

Declination, to correct the. One of the processes of working up the sight, at noon, after which a drink is in order. See **Yardarm.**

Deepwater. When pronounced with the accent on the first syllable, this is an adjective, a déepwater voyage. If an added accent is laid on the middle syllable, it is an adverb, "he always went déepwáter."

Demurrage. The penalty in admiralty law for delay beyond the stipulated lay days by consignees in removing cargo. Sometimes used figuratively alongshore.

Departure. A term in navigation, meaning the base line from which begins the daily reckoning of courses and distances sailed. See page 3.

To take one's departure, meaning to leave a place, is, of course, standard English and owes nothing to the sea. But when a person, in argument, says "I take my departure from"—certain basic facts, and then goes on to posit certain conclusions, he is employing the language of the navigator.

Derelict. A ship abandoned but still afloat and a menace to navigation. A term in admiralty law, from the Latin derelictus, meaning one who is unfaithful, neglectful of duty or obligation. Its colloquial shore use to mean a person who is broken down by a vicious life, carries more of the nautical meaning than of the standard classical use. See **Castaway.**

Dévil in a gale of wind, busy as the. A self-explanatory phrase, common alongshore. "I don't care if you are busy as the devil in a gale of wind; I got more to do myself than a man on the town [a pauper]." For the devil to pay, see **Hell to pay.** See also betwixt the devil and the deep blue **sea.**

Dido, to cut a. Now completely a shore phrase; said to come from H.M.S. *Dido*, a very fast ship, whose commander used to sail her in circles around other vessels of his squadron to show off her fleetness.

Different ships, different fashions; some carry their jib-booms aft. A sardonic comment current at sea and alongshore, when something strikes the speaker as perfectly outlandish.

Discharge. (1) (With accent on the first syllable.) The seaman's record of service. Alongshore, it may mean a recommendation, certificate, to-whom-it-may-concern letter.

(2) (With accent on the last syllable.) To take cargo or ballast out of the ship. It is used similarly alongshore. "Wait till I discharge this load of coal."

The shore use—to discharge an obligation, duty, debt, and so on—probably precedes this specialized nautical use.

Distress. A ship in distress is one in need of assistance, indicated through various flag **signals** by day, and by rockets and "blue-lights" at night (as well as, nowadays, by radio).

"To fly signals of distress" is occasionally met with in literature. See **S O S, Union down.**

Ditty Bag or **Box.** A sailor's "housewife," or kit of small mending gear. The term is used in the same sense alongshore.

Dock (Dutch). A basin between wharves for the reception of vessels. The dock in which accused persons appear in court is believed by etymologists to come in some obscure way from the marine dock.

Doctor. See **Forms of address.**

Dog's-body. A sea dish of peas boiled in a cloth like pudding.

Dogwatch. The period from 4 to 8 p.m., which is divided into two watches in order that periods of night duty may alternate daily. See **Time, Watch.** In newspaper slang, the dogwatch covers the evening hours when the paper is being made up.

Doldrums. A belt of calm, rainy weather at the equator, separating the *Trades.* By landspeople, the word is used to mean low spirits, ennui, "the dumps," and this usage was probably the earlier.

Donkey's breakfast. A straw mattress in a sailor's bunk. The term is used jokingly and disparagingly alongshore.

Double-Dutch. Unintelligible language. According to Bowen, this was originally a seaman's phrase, referring to a method of coiling rope. The complete phrase was "double-dutch and coiled against the sun."

Double-tides. To work double-tides is to have a long-drawn-out spell of work, originally one that could not be completed in a single ebb-tide. "We worked double-tides hauling up the boats."

Douse. Sailor's slang, meaning to abate or extinguish. To douse the topsails is to furl them; to douse the glim is to put out a lamp or lantern. The latter phrase is still used jocularly by shore people.

Alongshore, douse is a more emphatic word than pour. "Douse a bucket o' water over it, quick!" "Don't douse so much cream on your strawb'ries; you'll drown 'em." Whether this comes from the nautical use of the word has been questioned.

Down East. A general term for Maine and the Maritime Provinces of Canada. People from these parts "go up" to Boston, and return "down home." A "down-easter" may be either a person or a vessel hailing from that region.

Down the hatch. See **Hatch.**

Dragnet. A net used by fishermen for bottom-fishing. It is police and newspaper slang for a round-up of suspects.

Drag water aft. Same as broad in the **beam,** q.v.

Draw-bucket. A canvas bag about 2½ feet long, used to dip up water from over the side. Alongshore, since I was the bigness of a draw-bucket means since I was a small child. See **Trawl-keg.**

Dredge (Pronounced drudge). To scrape up from the bottom, as shellfish. All land uses of the word come from the sea term, except the culinary one—to dredge with flour, etc., which is from another source.

Dress ship. See **Signals.**

Drift. (1) The rate of speed when drifting or hove to. See **Heave to.**

(2) Calculation for the influence of **current**. The land phrase, to get the drift of what someone is saying, is probably from this source.

Driver, a. A shipmaster given to carrying sail. Alongshore, it means a fast worker. See **Carry on.**

Drogher. A disparaging name for an old vessel in some disagreeable trade—lumber-drogher, "dungdrogher" (one bringing guano from the west coast of South America), etc.

Drydock. A place for repairing ships. Alongshore, it is a joking name for hospital.

Dry nurse. An experienced chief mate serving under an inexperienced or incompetent captain. For the latter, the British seaman had a name, too: "Paper Jack," implying that he held his job through influence.

Dry-skin. A whale with scarcely any blubber. Hence, in whaling communities, a man who leaves less property than was expected. (Cf. the British expression "he didn't cut up well.")

Duck (Dutch doek, German tuch, cloth). **Canvas,** but particularly the light variety used for small sails, sea bags, hammocks, and sailors' warm-weather clothing. Similar goods are currently sold under this name ashore for garments and upholstery.

Due. Exactly (used at sea as an intensifier of the four cardinal compass points: "the wind was due east"). Due in this sense is merely an extension of the word in its shore sense of proper. See **Dead.**

Duff. A boiled or steamed pudding served on Sundays at sea. It is known by this name alongshore, and is not unknown to landspeople. The story goes that an Irish cook found a recipe for dough pudding, and tried it on the crew. "What d'ye call it, Doctor?" he was asked. "Duff—here it is in the book," he answered. "But that's dough," objected one erudite seaman. "If R-O-U-G-H spells ruff, and T-O-U-G-H spells tuff, why the hell don't D-O-U-G-H spell duff?" was the cook's silencer; and duff it has remained.

Dungarees. Working-clothes of blue jean. The word is of Hindustani origin, first applied to sailors' clothes, and by them transplanted ashore.

Dunnage. Loose material laid under ballast or cargo to protect the ship's skin. Sailors adopted the word to mean clothing, personal effects; and coastal people still use it in this sense.

Dutch, beat the. "If that don't beat the Dutch!" is a phrase common among landspeople. It comes from the battles between the navies of Britain and Holland in the 17th century, in which the English sailors found the Hollanders unexpectedly doughty antagonists.

Dutch courage. Liquor; from the gin which sailors believed was always served out in the Dutch navy before a battle.

Dutchman. On shipboard, a "Dutchman" is more often a German than a native of the Netherlands. The phrase common ashore: "or I'm a Dutchman," i.e., don't know what I'm talking about, comes from this somewhat disparaging old sea term. See **Flying Dutchman.**

Dutchman's anchor. Something important that has been forgotten or left behind; from the old jest about a Dutch shipmaster who had forgotten to bring his anchor along, and so lost his ship.

Dutchman's breeches. A small patch of blue sky at the end of a storm is spoken of as "enough to make a Dutchman a pair of breeches."

Dying man's dinner. A bite to eat, snatched in heavy weather, or when the ship is in difficulties and no regular meals can be served; or, alongshore, when housecleaning is going on.

Ease off or away. To steer less closely into the wind; let out more **sheet.** Ashore, the phrase means to be less severe; moderate one's conduct.

East about. To go east about is to leave the China Sea by one of its eastern passages, so-called, and not through Sunda Strait, the southern and more usual exit. No other compass point has about appended to it in this way, nor does the seaman apply the phrase to any other region of the seas.

Easting. Progress to the eastward; the ship makes easting. To run one's easting down is, specifically, to round the Cape of Good Hope eastbound.

East Injies. Pronunciation of East Indies.

Eastward (Pronounced east'ard). Towards the east.

Eat hearty, and give the ship a good name. A hospitable prompting to guests, often heard at 'longshore tables.

Ebb. The outgoing tide. For leeward ebb, windward ebb, see **Tide.** See also **Current, Flood, Water.**

In shore language, at a low ebb means in poor condition or straitened circumstances. To say that one's life is ebbing away, or, more poetically, that one is "going out with the ebb tide," means that one is dying.

Embargo (Spanish, meaning a restraint or prohibition). A governmental order preventing ships from leaving port. Ashore, it is figuratively used in various connections; to put an embargo upon things or ideas.

Embark upon. To take ship. Now mostly a shore phrase, meaning to start an enterprise. Disembark is, however, never used figuratively.

Encompass, to. See **Compass.**

End for end. The other way around. Its use seems largely confined to shipboard and alongshore.

Ensign. See **Colors, Signals.**

Even keel, on. See **Keel.**

Everybody's mess. A sea and coastal term to describe a busy-body, a kibitzer, is "in everybody's **mess** and nobody's **watch.**"
"Every hair a rope-yarn and every drop of his blood Stockholm tar." A seaman's description of a thorough seaman.
Every man Jack. See **Jack.**
Eye. See **Pipe one's eye, Weather, Wind, Windward.**
Eyelids, hanging or **holding on by the.** The seaman's vivid description of his situation aloft during a heavy gale. It has been adopted by landspeople to characterize any precarious situation.
Eye of the storm. The central calm in a cyclical storm.

Fag end. The frayed-out end of a rope. Ashore, it is a disparaging term for remnant, while fagged out means exhausted or dishevelled. The shore use may have been the earlier.

Fairway. A straight course down a channel; used similarly in golf.

Fair-weather sailor. A shore expression meaning someone who cannot be depended upon when the going gets hard. (Cf. fair-weather friend.)

Fair wind. See **Wind** (2).

False color. See **Colors.**

Fare. (1) A fisherman's catch of fish. Alongshore, a full fare may mean a bountiful crop, or gains in general. Shore phrases, such as to "fare well" and "to live on the best of fare," may possibly come from this source.

(2) To fare also means to journey: fare far afield, wayfarer, seafarer, etc. In this general sense, it is a land, not a sea term. But in the course of centuries, the traveller himself, and the money he pays for transportation, have come to be known as fares. It is in this sense, and directly from seafaring, I believe, that the railroads have taken over these terms.

Fast and loose, to play. A shore phrase, by some supposed to be of nautical origin (referring to making fast and loosing sails), but more probably it comes from an ancient game played with a string or strap.

Fathom. (1) The shortest measure of distance at sea; six feet, originally the measure of the outstretched arms. See **Cable's length, Knot, Mile.** Alongshore, it is used as a rough measurement of rope, etc.

Shakespeare speaks of drinking "healths five-fathom deep."

(2) To ascertain the depth of water between the ship and the bottom, in fathoms. As a verb, it is obsolete; the nautical term today is "to **sound.**" It is still in use ashore, however: "His reasons were hard to fathom." See **Lead, Plumb.**

Favoring wind. See **Wind** (2).

Fender. A buffer of wood or braided rope hung over a vessel's sides to prevent injury to her hull from contact. The name for the fender of an automobile is probably taken from this nautical source, even though the verb, to fend off, may have had earlier general use ashore.

In nautical language, to fend off is to push with an oar or barge-pole. A coastal expression is "Hardest fend off!" meaning something like "May the best man win!"

Fetch, Fetch by. To succeed in weathering a point of land. Ashore, it means to attain an objective. "He couldn't quite fetch (it)." To fetch the turf is a fisherman's term for reaching port.

Fiddler's Green. A British sailors' term for that fine place where souls of good sailors go after their bodies go to **Davy Jones' Locker.**

Figurehead. A carving at a ship's bow. Alongshore, it is facetiously used for face. "He's got a figgerhead that would stop a clock!" In shore speech, it denotes a person holding a sinecure, one whose name is used for display purposes: "Nothing but a figurehead!"

Filibuster. To run contraband of war to revolutionaries. Ashore, it is a political term, meaning to hold the floor to prevent a bill coming to a vote in a legislative assembly. The word in its earliest form was probably the Dutch vrijbuiter (freebooter), a sea robber or pirate. See **Piracy.**

Fill away. To keep away from the wind till the sails are full and drawing. Alongshore, to fill away is to make haste.

Finn. See **Squarehead.**

Fin out. When a whale rolls over to die, he is said to be fin out. In whaling communities the term means moribund. "Old Cap'n Peleg is nigh about fin out."

Fire escape. See **Sky pilot.**

Fire-ship. In naval warfare, a vessel loaded with combustibles and maneuvered so as to fall among enemy shipping. A waterfront prostitute believed to be infected with venereal disease is called by this apt name.

First-chop. Of first-class quality; from the Chinese merchant's chop-mark. Pidgin English, brought home by sailors.

First-rate. An official rating of warships in the British Navy of old, but now taken over into shore speech with no recollection of its nautical origin. See **A 1.**

Fish. (1) This word appears in many shore expressions, for example, to fish (around) for information or news; to fish for compliments; to fish in troubled waters that is, to take advantage of the misfortunes of others. "All's fish that comes to his net" is said of a person who contrives to turn everything to his own advantage, or who is unscrupulous. "Speak your mind if you never sell a fish," is sly advice not to be too outspoken. "A pretty kittle of fish" alludes to a predicament or mishap. Shore users sometimes say "kettle," which is a mistake; the word is properly kiddle or keddle, an old English term for a fish weir. See **Blow your horn,** fish or cut **bait, Trip** o' fish.

(2) To fish also means at sea, to mend a spar by setting in a brace or splint. It is applied, too, to one of the processes of getting the anchor fast on the bow after weighing.

Alongshore, the word is used for mending jobs. "I want you should fish that clothes-horse for me today." In shore speech, the device holding the ends of railroad rails together is a fish-plate. These are probably all related uses, and are thought to come from the French fiche, a peg. Professor Chase feels some doubt about this derivation, because of lack of evidence that this word was used by French sailors, and in the general sense of fastener.

Flag(s). See **Colors, Signals.**

Flakes, on the. Flakes are frames on which fishermen lay split and salted fish to dry. "On the flakes" is a fisherman's slang term for dead, "laid out."

Flash packet. A name for a fast, showy vessel; and inevitably transferred to a woman of the ports to whom those adjectives could also be applied. (The slang sense of packet as a case of venereal disease may have crept in here as well.)

Fleet. (1) A group of ships in company, or belonging to the same owners. Ashore, it has come to be used in the same sense of cars, trucks, airplanes, etc.

(2) To **overhaul** a **tackle;** to move or slide along cautiously. Alongshore, it is used in the latter sense. "Fleet forrard, folks;

there's too much cargo aft," said the deacon who wanted the front benches filled at prayer meeting. The only sense in which this remains in standard English is in the adjective fleeting—i.e., transitory.

Flip. A mixed drink, familiar to landspeople. It is apparently a sailor's word; an early writer calls it "a sea-drink of small-beer (chiefly) sweeten'd and spiced upon occasion."

Float. This word is used ashore in a special sense, meaning to organize and sell shares in a company or business, to float a loan. See **Adrift, Afloat.**

Flood. The incoming tide. For leeward flood and windward flood, see **Tide.** The word flood used to designate the sea itself is purely literary and of the land. See also **Ebb, Water.**

Flotsam. Goods found floating on the sea; admiralty law defines their ownership. The word is used poetically and somewhat affectedly in literature, often combined with **jetsam,** to mean stray gatherings, especially literary miscellany. (The third of the trio, lagan, or ligan, goods sunk and buoyed to be later retrieved, has never appeared in shore speech.)

Fluking, all a-. Sailing fast and furiously. A whaleman's expression, from the speed of a whale driven by strokes of its powerful flukes, or tail-fins. The whale's flukes were named from their resemblance to the flukes of the old-fashioned anchor.

The phrase has passed into shore speech, to mean rapid, unimpeded progress of any sort.

Flying colors. See **Colors.**

Flying Dutchman. A person dogged by ill-luck; from the legendary Dutch sea captain, Van der Decken, condemned for impiety to cruise forever off Cape of Good Hope. The tale is now embedded in world folklore.

Flying-fish weather. Fine, sunny weather with gentle winds, as found in the tropic seas where flying-fish abound.

Fly, let her! A colloquial shore phrase of encouragement, which probably comes from the sailor's exclamation of delight when the ship was booming along. A ship empty of cargo is said to be "flying light." (To let fly is, however, not of sea origin, but probably comes from archery or hawking.)

Fly the blue pigeon, to. Sea-slang for casting the hand lead is to "fly the blue pigeon," but this phrase, although it appears in most lists of sea terms, was not originated by sailors. In the late 1700's, it was London thieves' slang for robbing metal-roofed churches of the sheet lead used to cover them. It is not found in the sailor's vocabulary earlier than 1870. See also **Sound.**

Foam, the. A poetical, landsman's term for the sea.

Fo'c's'le. Pronunciation of forecastle, the sailors' quarters in the forward house.

Fog. A peculiar and dreaded danger at sea, though not upon the land (at least up to the time of automobile transportation). Hence, the figurative shore phrases in a fog, befogged, meaning confused, puzzled, probably hark back to the seafarer's experience. See **Thick.**

Foghorn. The mouth-blown tin foghorns on sailing-vessels, which later gave way to hand-operated foghorns, emitted a raucous blare; even more noisy and plangent are the shore signals and whistling buoys that warn ships of dangers they cannot see in a fog. Even inland people speak of a "voice like a foghorn"; to the coastal speaker, a person with a loud, hoarse voice "bellers like a foghorn."

Fog mull. Several days of intermittent fog without wind.

Following wind. See **Wind** (2).

Foo-paw (French faux pas). A whaler's term, meaning a bungling job in killing a whale, the expression having been picked up from French whalers met at sea. It is common in whaling communities; but the "forks pass" of the landsman is probably an educated quip by those who have studied French.

Fore-and-aft. Lengthwise of the hull; the opposite of athwartships. It is also the name for rigs in which the sails are set lengthwise of the hull; the opposite of **square-rigged.** Applied to a person, "fore-and-after" means a pervert, but no such significance attaches to **schooner**-rigged, q.v.

Alongshore, fore-and-aft means the whole length of anything. According to Chase, the main streets in Havana, following the shoreline, are said to run "four-and-a-half," a corruption of fore-and-aft.

Forehanded. See **Hand.**

Foremast hand. See **Before the mast.**

Forereach. To shoot ahead of another vessel. In shore speech, to get the better of another person.

Fore royal, the. The early-morning mug of coffee on shipboard; the first cup alongshore. Just why it should be called after the topmost sail on the foremast is not known.

Foretop. The platform at the junction of the foremast and foretopmast. By landsmen, this word is sometimes used to mean a person's forelock, or the front part of a horse's mane.

Forge ahead (Probably a corruption of "force"). To gather headway, or to outsail another vessel. In shore speech, it means to force one's way onward, distance competitors.

Forms of address. The captain of a merchant vessel should always be addressed by his title. By his subordinates and business associates, he may be addressed just as "Cap'n," or as "Cap'n Smith," or whatever may be his surname. By his familiars and young friends, he may be called "Cap'n Charles," or even "Cap'n Charlie," if that happens to be his first name. But after he attains the rank of captain, few except his relatives and boyhood friends would venture to use his first name without the "handle."

To address him as "skipper" is a social error of the first order—only the masters of barges and harbor craft answer willingly to this title. **Old Man** is of course a title of reference and not of address.

Master and commander are ratings and are never used as titles of address in the merchant marine. (Though master used to be the only proper term. A sea dictionary of 1708 says: "Captain is the commander-in-chief aboard any ship of war; for those in merchants are improperly called so, as having no commissions and being only masters." This pretty closely fixes the time when the change-over from master to captain of merchantmen took place.)

In replying to an order or question from the captain, neither mates nor men ever omit the word "sir." Plain yes or no would be tantamount to insubordination. Even in general conversation, as at the table, mates are careful to "sir" the captain.

You will often hear a captain, in addressing another shipmaster with whom he is not well acquainted, a passenger, or a shore person of any importance, also use the word "sir." In this case, it connotes not so much respect, as a dignified formality. Shore people, both men and women, will do well to bear this in mind, and to employ the word in speaking to captains. It is surprising what a change in atmosphere that simple word will often bring about. One can almost hear the shipmaster say to himself: "Ah! *here's* somebody who understands shipboard etiquette!"

On large vessels, the chief officer and the chief engineer may properly be addressed simply as Chief. Junior mates and engineer officers are addressed either as plain "Mister," or as Mr. followed by the surname. They must be "sirred" by their inferiors when on duty, but off duty, their relations among themselves are less formal than those obtaining among captains.

The ship's cook answers to "Doctor," the carpenter to "Chips," and the sailmaker, if there are any left, to "Sails." The boatswain can be abbreviated to "Bose" without injury to his dignity. The men of the crew are addressed either by their first or last names, without prefix.

The proper acknowledgment of any order from a superior is **"Aye, aye, sir!"** The answer made by an officer to a report from

an underling is likely to be "Well" (an abbreviation of the old British naval term "Very well done"). "Well the starboard braces," etc., is commendatory dismissal—things are the way the officer wants them. In the case of certain reports, the standard answer is "Make it so." See **Make** (3).

On naval vessels, the conventions regulating all these matters include the etiquette of hailing and saluting, and are too complicated to permit their being described in detail here.

Forward (Pronounced forrard). Toward the bow; in front, but on board (forward and outboard is ahead). Alongshore, it is used in the same sense, and also figuratively: "I can't seem to get any forrarder with this job."

Foul. Bad, as of the weather; entangled, as of the anchor, ropes, vessels in collision, etc. In both these senses, foul is the opposite of clear. See **Afoul,** foul **hawse.**

Founder. Of a ship, to fill with water and sink. Of animals, to bloat and become ill from overeating or drinking. This land use may have been the earlier, but as founder is used in shore speech today, meaning to come to complete ruin, it would seem to be rather the sea use that is followed. See **Swamp.**

Freeboard. That portion of a ship's side showing above the water line. Ashore, it means height above some base line. See **Board.**

Freight. Goods transported by ship; the term later came to mean payment for such transport. Cotgrove, writing in 1611, mentions "the fraught, or fraight, of a ship, also the hire that's payed for a ship or for the fraught thereof." The word "fraught" is now used only in poetry.

Freight is now generally applied ashore to goods sent by railroad and to the cost of their transportation—the latter being shortened from freight money. The slang phrase pay the freight probably comes from this source, and is of secondary derivation from the sea.

Frenchy. See **Dago.**

Freshen. (1) Same as **breezen,** q.v.

(2) To shift a rope so as to bring the wear in another spot. Figuratively, to freshen one's **hawse** is to take a small drink.

Frigate. An old-time sailing ship of war. The frigate bird of the tropics belongs to a species christened Fregata in honor of these old vessels. Frigate-built means heavy, obese—said principally of women's figures. See **Clipper, Man-o'-war** bird.

Frisco. The sailor's name for San Francisco. San Franciscans don't like it.

Full steam ahead! A term developed in steamers from full sail ahead. Ashore, it is used to mean with might and main. It is almost the only phrase which the steamship has contributed to shore speech.

Furl. To take in a sail and tie it with **gaskets** around a **spar.** The word appears in shore speech in connection with umbrellas.

Gaff. (1) The spar at the top of a **fore and aft** sail. Most lists of nautical expressions carry "to blow the gaff" (give away a secret) and "stow your gaff!" (quit talking, shut up) as derived in some forgotten way from this spar; but etymologists believe that they come from a different gaff—a name for a public fair in old England.

(2) A gaff is also a pole with a hook in the end, used to lift fish on board a fishing-vessel. This is the source of a 'longshore verb, to gaft, or sometimes to gaffle on to something. "I'm goin' to gaffle on to that last piece o' punkin pie."

Gale. A gale **o' wind** is a heavy storm; a livin' gale, a wind of hurricane force, at sea and alongshore. Shore use of the word to mean a gentle, fair wind—a favoring gale—is now purely literary. See **Blow, Breeze, Devil.**

Galleon (Spanish). A large **galley;** a type of vessel that has been obsolete for several centuries. Use of the word by moderns is literary and poetic.

Galley. (1) An old-time vessel, originating in the Mediterranean and usually propelled by oars. The galleys of the Barbary corsairs were rowed by captives and criminals, under conditions of incredible hardship. Hence, in shore speech, we have "I'm nothing but a galley-slave" (though when this is said by a housewife or cook, some of the meaning of galley (2) may have crept in also.) Printers refer to themselves as galley-slaves because of the long type-trays called galleys with which they work.

The sophisticated quip "What are you doing in that gallery?"— i.e., among those odd people—arises from a mistake in translating the French phrase "dans cette galère," which means a galley, not a gallery.

Galleyworm is a name for the common milliped given because of the fancied resemblance of the creature's many feet to the oars of a galley. Gallipots, the crockery jars used by apothecaries, got their name because they were first brought to

England in the later commercial sailing-galleys of the Italians. See **Galligaskins.**

(2) A ship's kitchen (formerly called **caboose,** as in Dutch vessels), probably acquired the name of galley from galley (1) in some obscure way. The galley shares with the cook the status of a stock-in-trade for jokes among the crew. Galley news, or galley yarns, are the rumors that float about on shipboard. The "galley downhaul" is an imaginary rope which **greenhorns** are told to find and let go.

"Galley stove, to take everything but the." The shore equivalent is, of course, everything but the kitchen stove. It would be difficult to find out which came first.

Galley-west, to knock. See **Knock down.**

Gallied (From old English gallow, to frighten). The whaleman's term for the behavior of a whale after being struck. Alongshore, it means confused, embarrassed.

Galligaskins. Wide trousers, also called "shipmen's hose" and "Venitians," worn by English seamen in the 16th century. Webster's Dictionary gives an involved derivation from "Gallic Gascons"; but the alternate name "Venitians" makes it fairly certain that they got their name from the Italian galleys in which products from the Mediterranean reached early England. See "gallipot" under **Galley** (1).

Galligaskins were worn by shore people in the Dutch settlements about New York, as described in Washington Irving's *Rip van Winkle.*

Gam. Among whalemen, to visit back and forth between ships at sea is to gam; hence, alongshore, a long, friendly chat. "Come over after supper and have a gam."

(The cross-word puzzle definition of gam as a school of whales is a complete error; the whaleman's term for this is pod.)

Gangway. "Originally in general sense, then nautical." (Weekley) A passage along a ship's deck, also the opening in the bulwarks through which one goes aboard her, and the side-ladder for climbing up her side—all are called gangways. In shore speech, any passage way for foot traffic may be called by this name. "Gangway!" or "Get out of the gangway!" is a call to get out from under foot.

Gasket (Earlier, gasset). A lashing used in furling sails. The word is jokingly used alongshore to mean a sash or belt. (The shore term meaning packing ring or sheet is apparently not connected.)

Gear. Rigging, especially **ropes,** blocks and **tackles.** Alongshore, it means clutter. "Get your gear off that table so I can get it set for supper." The origin of the word is uncertain, and its land use may have been the earlier.

Gilguy. The sailor's disparaging term for an improvised and inefficient bit of rigging. A description of a woman's headdress, found in an old sea journal, is "an odd affair of gilguys and braces."

Gills, hanging by the. A fisherman's description of a fish improperly hooked, and likely to be lost. Alongshore, it is said of anything in a precarious state. "The children have been swinging on the front-yard gate till it's only just hanging by the gills."

Gimbals. Concentric rings to keep the compass, lamps, and other objects, always upright at sea. Alongshore, gimbal-jawed means having a protruding or wabbling lower jaw.

Gingerbread-work. Gilded carving and scrollwork used to adorn a ship; from the gilded gingerbread sold at British country fairs. In shore speech it means excessive ornamentation, particularly architectural.

Girls have got hold of the towrope, the. The sailor's exclamation when all is going well, and the ship proceeding fast and steadily towards home.

Glass. In the old days at sea, a glass was half-an-hour; from the sandglass with which time was measured.

> He chased us to windward for glasses one or two;
> He chased us to leeward, but nothing could he do.

To later seamen, and to coastal people, the glass means either the barometer or the old-fashioned telescope, while glasses are binoculars. "The glass is falling steadily; we're in for a breeze o' wind." "I made her out with the glasses, coming up between the heads."

Glory hole. The strong room on old-time ships, where specie and treasure were kept. On modern steamships, it is a slang term for the living-quarters of the cooks and stewards. Alongshore, it means a locker for odds and ends, a disorderly one being implied.

Godown. A warehouse or storehouse for goods. Pidgin English, corruption of Malay, godon, warehouse, brought home by sailors.

Go into steam. To leave sailing-vessels and take employment on steamships—sarcastically elaborated into "leave the sea and go into steam." See **Windjammer.**

Goney, Gooney. The sailor's name for an awkward sea bird found in southern latitudes. Alongshore, a lubberly, though not necessarily a "lacking" person.

Govern. From the Latin gubernare, meaning to steer a ship. A gubernator, from which comes our modern governor, was originally a ship's helmsman.

Grampus. The blackfish, a species of small whale often met at sea and seen along the coast. To puff like a grampus is to pant heavily. To blow (or tip) the grampus is to pour a bucket of water over a man found asleep and snoring during his watch on deck.

Granny. Shortened form of a granny's knot, a square-knot improperly tied so that it will slip. The mark of a landlubber. See **Reef knot.**

Grapnel, Grapple. To hook something up from the bottom, or to fasten to the rigging of another ship when boarding her in battle, by means of a grapnel or multiple-armed hook. All land uses of this word are derived from the sea use, whether grasping with the hands or going through mental struggles is implied.

Shakespeare, who never makes a mistake in his nautical metaphors, advises:

> Those friends thou hast, and their adoption tried,
> Grapple them to thy soul with hooks of steel.

"Hooks" appears in only one edition of Shakespeare; we may be sure that the editor who changed it to the more familiar "hoops" was a landlubber. (It would be a fascinating job for someone with a nautical background to do a doctor's dissertation on Shakespeare's use of sea terms. It might be possible to prove that the Bard had been a sailor once!)

Grass about the bows. Alongshore, a facetious term for whiskers.

Grating. An openwork grid of wood or metal strips, used to cover a hatch for ventilation, or as a foothold or protection against wet decks. "He can't see a hole through a grating" indicates either stupidity or drunkenness. (Probably not originally a sea word.)

Gravelled, to be. To be stumped. It refers to the stranding of a vessel.

Gravy-eye watch. The middle watch, from 12 to 4 at night, when the eyes felt sticky. Modern sailors have garbled it to graveyard watch, whence the graveyard shift in industrial plants. See **Time.**

Greasy luck. The whaleman's farewell, used in whaling communities whenever one would say "Good luck!"

Greenhorn. Although this word seems to belong to the sea—a greenhorn or "greeny" is a person making his first voyage—its very etymology shows that it must have originally been a shore term for a young horned animal.

Greyslick. See **Slick.**

Grind one's own bait. See **Bait.**

Grog. Rum and water, first served out to British seamen serving under Admiral Vernon—before that, they got the rum neat. Because of his grogram (grosgrain) coat, he was known as "Old Grog."

Among shore people, a mixed drink is called grog; a grogshop is a low kind of liquor saloon; a grog blossom is the red nose of a chronic alcoholic; while groggy means dizzy and staggering.

Ground. To run a vessel **ashore** or **aground.** An equivalent phrase is to take the ground. The landsman's term, to run into the ground, i.e., to carry a thing too far, to dull by repetition, is probably from this source.

Growl. To find fault, complain; a sailor's term for anything from a mild mutter to an angry roar. "Growl you may—but go you must!" expresses the seaman's philosophy.

Grub. The sailor's term for food in general. Some etymologists say the word originated on the land; but how can they be so sure of this concerning a coined word, which goes back to no known roots? The fact that it may appear first in shore literature is neither here nor there—sailors produce little literature. See **Skoff'm, Manavelins.**

Gudgeons. The hinged sockets into which the hooks, called pintles, of the rudder are inserted, thus permitting it to swing freely. The word was originally **googing** (thus spelled by Falconer). It came from the French gouge, English googe, a name for a waterfront drab. The etymology of pintle can easily be inferred. But the obscene implication has long since been forgotten, and gudgeon and pintle are now innocent and respectable bits of marine hardware.

Gundalow (Pronounced gunlow). A type of barge or scow built along the New England coast for carrying salt-marsh hay and other cargo. The name is probably from the Italian gondola, a word brought home from the Mediterranean by sailors.

Gunk-hole. A Maine coast term for a narrow, rocky cove, difficult and dangerous to get into and out of.

Gunny-bag, Gunny-sack. From the East Indian gunni, a coarse cloth made of jute, brought home by sailors.

Guns. To stick to one's guns is to maintain one's position in an argument—a phrase going back to naval warfare rather than fighting on the land. See also to **spike** one's guns. "Great guns!" a meaningless exclamation of surprise, is probably a contraction of **blow** great guns, q.v.

Gunwale (Pronounced gunnel). This originally meant the top **strakes** of a ship's planking, but with the passage of the wooden ships, it now means only the rim around a boat's sides. A common phrase alongshore to describe a great plenty of something is gunnel-full and running over. Loaded to the gunwales is occasionally heard in shore speech, of an intoxicated person, or an over-full conveyance.

Gurge (From the old English gurge, a whirlpool). A fisherman's term—a boat being driven under an excessive amount of sail is said to gurge along.

Gurry. Refuse and slime from cleaning fish—hence, alongshore, disgusting dirt of any sort. "Gurry-butt" is a vulgar name for garbage pail.

Guy. Same as **stay,** q.v.

Hail. (1) To call out from one part of the ship to another, or from ship to ship. See **Call.** Alongshore, it may mean to call up by telephone: "Give me a hail when you're ready to start." Its shore use, meaning to welcome or acclaim, "hail the dawn of a new era," may have preceded its use at sea.

"Within hailing distance" means near enough to call from ship to ship. The phrase appears frequently in shore speech: "He now felt that he was within hailing distance of success."

(2) A ship hails from her port of registry. A landsman is often said to hail from a place, though not necessarily his birthplace.

Half-mast. See **Colors.**

Half-seas over. Partly intoxicated, a sea term in common use ashore.

Halibut. The common food fish, properly pronounced by coastal speakers hollibut. It was originally "holy-but," a fish to be eaten on Fridays. Landspeople generally make the first syllable rhyme with "pal."

Hammock. The name and the article itself were very early adopted by sailors from the Carib Indians of the West Indies, and were first used by Europeans on shipboard.

Hand. (1) A man before the mast; a common sailor, described as green hand, old hand, and by other terms. The common use ashore of hand to mean a workman, had its origin at sea. All hands means the whole ship's company. (The nautical superlative is "All hands and the cook.") All hands has likewise come ashore, to denote everybody concerned, or the whole party. Shorthanded, common at sea and ashore, has the same derivation. Forehanded, meaning prudent, saving, is probably of nautical origin.

A master hand, a colloquial phrase meaning a skillful person, and a good hand at something, come from sea usage. "One hand for the ship and the other for yourself" is sardonic advice alongshore not to be over-forward in the interests of one's employer.

"Bear a hand!" meaning "hurry up," and "lend a hand," or "come and help with the work going on," orders given at sea, are both common in shore speech. "All hands to the pumps!" is a call for concentrated effort.

(2) To hand means to handle, especially ropes and sails. "Able to hand, reef and steer" were the stated qualifications for the rank of **ordinary seaman.** By extension, alongshore, this phrase is used as qualified praise, for a person who can "get by" acceptably, but no more. "Oh, I guess he's able to hand, reef and steer."

Hand-a-running, Hand-over-hand, Hand-over-fist. Fast and easily, as of a rope that can be hauled in readily without taking a turn around a belaying-pin.

Alongshore, hand-a-running means in succession. "I've drawn a spade three times hand-a-running." But hand-over-fist has rather the meaning of rapidly, without exertion: "The money's coming in hand-over-fist."

Handsomely. Occasionally used alongshore in its original sea meaning of slowly and carefully.

Handspike. A tool used on board ship. See **Spike.** Alongshore, stiff as a handspike means unconscious or dead; "fits like a purser's shirt on a handspike" signifies baggy, too loose. (The purser on naval vessels was second only to the cook as the object of rude remarks. The shirt referred to is the cheap stuff he sold to the crew, not his own garment.)

Hannah Cook, don't amount to. A coastal phrase of disparagement. Said to be a corruption of a hand and a cook—the crew of a very small coasting vessel.

Harbor. A sheltered port for ships. Weekley says it originally meant shelter in general, "but naut. appears early." Ashore it is used figuratively, "a snug harbor in his old age," etc. For landlocked, see **Land** (1). By extension it becomes a verb, to harbor, meaning to shelter, entertain—but usually with an unpleasant connotation (to harbor insects, a grudge, etc.).

Harbor-stow. An extra finish in furling sails on entering port. Alongshore, this is praise for an especially neat or fancy job.

Hard bread, Hardtack. Ship's biscuit; terms in common use alongshore, and readily understood, at least, by landspeople.

Hard and fast. Of a ship, the superlative of **aground.**

In shore speech the term is used principally to qualify the phrase "rules and regulations."

Hard lines. See **Lines.**

Hard up. With the rudder over as far as it will go, so that nothing further can be done. In shore speech, it means penniless. See **Bitter end.**

Harpoon. The whaleman's harping-iron, with which the whale is "ironed" by the "harpooneer." A harpoon is darted, not thrown.

Alongshore, "put them biscuits within dartin' distance" means "so that I can harpoon one with my fork." To harpoon is sometimes used in this sense by shore people. For a speaker's use of it to mean sabotage, see page 4.

Hatch. The cover over a hatchway, or entrance to the ship's hold. Alongshore, a trapdoor, or, figuratively, the mouth or stomach. "He ate so much he couldn't close his hatch." The Navy toast "Down the hatch!" has the same significance, and has become current in shore slang.

In coastal dialect, under hatches means well stowed away (or sometimes, dead and buried); while to clap or lay under hatches is an old term for put in prison. To the hatches means full up: "The barn's jam full clean to the hatches." See **Booby hatch, Scuttle** (1).

Modern writers about the merchant marine are fond of the word hatch, which to them seems to be synonymous with hold. Their ships go about the seas "with their hatches full"—a dangerous condition indeed if they were to meet with heavy weather and ship a few green seas!

Haul (From the old English word "to hale," which now survives only in such phrases as "to hale into court."). (1) A catch of fish. Alongshore, a piece of good luck. "He made quite a haul at the poker game last night." See **Mainsail haul.**

(2) To move a vessel that is not under **way.** To haul her off is to move her away from shore or dock, into the **stream.**

Alongshore, to haul off is used as an intensifiying phrase: "I hauled off and let him have it." To haul off shore means to keep at a distance, discourage familiarity. To haul one's wind is to head closer to it; figuratively, it means to take a new line of action, or to avoid a quarrel. To haul out is to dock or beach a vessel for repairs (probably a survival from the time when vessels were actually hauled out of the water). Alongshore, it means to be sick, go to hospital. To haul up is the same as **lay** up, q.v.

(3) Of the wind, to haul, or haul round, is to change clockwise; to **veer.** The opposite of **backen** round.

Haulover. A bar over which small craft could be shoved when the tide served, so as to get from one cove to the next. Such a bit of half-submerged land is often known locally as "The Haulover."

Have got. Have, possess, "here it is." Pidgin English, brought home from China by sailors. The negative is no have got.

Haven. A sea term for harbor, obsolete except in place-names, such as Vineyard Haven, Milford Haven. Ashore, its use is purely literary, "the haven where he would be," and so on.

Hawse. The ship's bows, or the water between her and her anchor when it is down. **Athwart** the hawse means across, or even resting against, the bows. A **foul** hawse means that the anchor chains have become twisted around each other. Alongshore, "Don't you come athwart my hawse!" means don't interfere. See **Freshen** (2).

Hawse ahead. To move a ship at anchor by heaving in chain; hence, alongshore, to make slow progress; to inch along.

Hawsepipe (Hawsehole), through the. A shipmaster who began his seagoing life before the mast is said to have "come in through the hawsepipe"—i.e., the hole through which the anchor chain passes. Alongshore, it is a commendatory description of a self-made man. See through the cabin window.

Haze. To bully and knock about, in the manner of "hard-case" mates; and first used in this sense. College hazing got its name from the shipboard term.

Head. (1) The bow of a vessel. By extension, it meant in old-time sailing-vessels the foremast hands' latrine, which was under the forecastle head in the extreme eyes of the ship. "I've got to go to the head" is a euphemism alongshore, and is also used in the Navy of today, although the plumbing system to which it refers is quite different!

(2) Of a vessel, to head for is to steer or proceed on a definite

course. The standard query as to the course is "How's her head?"

The word is very common in shore speech: "I'm headed for home"; "headed straight for destruction," and so on. Quite possibly the colloquialism, "I told him where to head in at," i.e., put him in his place, comes from this source.

To be headed off means to be prevented by a change of wind from sailing in the direction desired. On the coast, "headed off shore" signifies getting well, out of danger.

Head, by the. Loaded so that the bow is deeper in the water than the stern. Alongshore, it is used in the same sense. "That porch looks down by the head a little."

Head-flaw. A sudden gust of wind from ahead. Alongshore, it means a bit of bad luck. Of some cocky person, one says: "It'll do him no harm to get a few head-flaws."

Headreach. To sail by the wind under shortened sail. It is sometimes used figuratively alongshore, meaning to make progress slowly or with difficulty.

Headway. Movement of a vessel forward through the water. A ship may lose headway, but she will have, or get, headway on, rather than make or gain headway. The latter are shore terms, meaning to make progress, usually against odds. See **Leeway, Sternway.**

Headwind. See **Wind** (2).

Health, a clean bill of. A certificate given by the port health officer to a departing vessel, facilitating her entrance to the next port. By landsmen, the phrase is generally applied to someone freed from blame.

Heave. At sea and alongshore, this word completely replaces throw or toss. "Heave us that plug of tobacco." "He hove away his chance." "She hove him overboard" (i.e., jilted him).

Many phrases of precise sea meaning employ heave. To get the anchor from the bottom is to heave it up. To heave ahead, or astern, is to move the ship in the direction indicated. See **Haul.** "Heave ahead!" as an exclamation, means continue, get on with it. "Heave and wake the dead!" is an injunction to put one's back into it. To heave short is to heave on the windlass until the anchor chain is perpendicular; a person on the point of departure is said to be "hove short." For heave in sight, see **Sight.**

In the sea use of the word, hove is the correct past tense. In the (probably earlier) land use, the past is heaved, "she heaved a sigh."

Heave down. To **careen** a ship in order to repair her or clean her bottom. Hove down is figuratively used alongshore to mean sick, or in the hospital.

Heave to. To stop the ship's progress by shortening sail and heading close into the wind, usually in heavy weather. See **Lay to.** Alongshore, "Heave to!" signifies "Stop, I want to speak to you."

Heel. To list under pressure of wind. Alongshore, anything out of the perpendicular may be heeled over. *See* **Careen, List.**

Hell afloat. A name for a hard-case ship, in which brutal treatment was the order of the day; used of quarrelsome households alongshore.

Hell or Halifax. Anywhere at all, so long as it isn't here! A phrase developed out of **coasting.** "Go to Hell or Halifax."

Hell and high water, through. A shore phrase, meaning in spite of every obstacle. It is probably of nautical origin. "Between hell and high water" is the same as "betwixt the devil and the deep blue **sea,**" q.v.

Hell to pay. The complete phrase is "Hell to pay and no pitch hot!" from the shipbuilding operation of filling the seams of the deck with hot pitch after caulking. It has been stated that "hell" was a name for the seam next to the waterway, because it was difficult to work on; but it seems more likely that a lively imagination conceived the difficulties of paying the seams of Hell itself.

The phrase is used by landsmen to denote an emergency not prepared for, and has been corrupted into the devil to pay on the false assumption that to **pay** involved the transfer of money.

Helm (Pronounced hellum). The rudder and the mechanism by which it is controlled. Use of the term at sea is limited to a few expressions. The ship minds (or does not mind) her helm. "Mind your hellum!" is a warning alongshore not to run into something or somebody. See also **Port** your helm, **Starboard** your helm. But in his ordinary speech, the seaman speaks of the **wheel.** See also **Steer, Course, Tiller.**

"At the helm" meaning in charge, at the head of, is a shore phrase entirely. See **Man** at the wheel.

Hen frigate. A vessel on which the master's wife and family accompanied him (with the implication that the "old woman" had too much to do with running affairs on board!)

Herm up. Of the weather, to become cloudy and threatening.

Herring. A popular food and bait fish, which enters into many shore phrases, "lose a sprat to catch a herring," for example. "Dead as a herring" is a fisherman's term, meaning very dead indeed, which has become common in shore speech. Another, less common, is "thin as a shotten herring"—one that has shed its roe. (The proverbial red herring, however, probably comes from hunting, not fishing.)

High and Dry. Aground too far to be lifted by the tide. In shore speech, left high and dry means deserted, powerless. See **Hard and fast.**

High line. A Banks-fisherman's term for the boat that brings back the best **fare**; used figuratively in shore speech for excellence in general.

High-water mark. The line left on wharf or rocks by the highest limit of the tide. To the landsman it means the height of perfection; the "top notch."

Alongshore, "high-water pants" are short trousers; while a boy who has failed to wash thoroughly is said to carry a high-water mark on his neck.

Hit between wind and water. Struck by a shell just above the water line. See **Hull.** Alongshore the phrase refers to an injury in one's mid-section.

Hit the deck. See **Deck.**

Hitch. (1) The distance covered on a single **tack**—same as a **board** or **leg.** The period of enlistment in the United States Navy is called a "hitch." Alongshore, the term is used figuratively: "I been goin' astern every hitch I made this year."

(2) The name of many knots used at sea and alongshore—clove hitch, timber hitch, a round turn and two half-hitches. Up to the hitches is a whaleman's expression for a harpoon driven in up to the point where the line is made fast to the

shaft. Alongshore, this is said of anything driven in deeply and firmly.

The landsman's use of the term—a hitch in the proceedings, an agreement, etc.—probably refers to an accidental round turn in a vessel's line, that jams in a block and prevents the rope from running through it.

As a verb, hitch does not appear in the seaman's vocabulary. The saying is that "a woman ties a horse, a farmer hitches it, and a sailor makes it fast." See **Make.**

Hogged. Said of a ship with her back broken, so that she shows a hump amidships; the opposite of **sagged.** It is used in the same sense alongshore. "The ridgepole of the barn is gettin' to look sort of hogged."

Hoist (Pronounced h'ist). To lift or raise with the hands, or by means of a **tackle.** The coastal pronunciation preserves that of the original old English word, hyse, with past tense hyst, the terminal "t" having become accidentally transferred to the present tense. The order to hoist, found in old songs, was "hoisa" or "hissa"; from this is said to come our somewhat obsolete cheer-word "Huzza!"

Alongshore, the word is sometimes used intransitively: "Please h'ist, Deacon Porter, so's I can get this chair past ye." It may even mean, contradictorily, a fall. "My feet went out from under me, and I got an awful h'ist."

Hold (Usually pronounced hole). (1) Cargo space in a vessel. The original word was, in fact, "hole," becoming "hold" through a mistaken etymology. "Hello-in-the-hole!" is a jocular greeting alongshore—to someone down cellar, for example. A swept hold expresses complete emptiness. "I'm ready to fall to anytime; I've come with a swep' hole," said the old fisherman invited to a meal. See **Hatch.**

In baseball, the man second from the batter in batting order is "in the hole" (hold). See **Deck.**

(2) (Always pronounced holt.) A hand-hold, generally on a rope. Used in the same sense alongshore. "Get a good holt on that basket." See **Ahold.**

(3) (Pronounced as spelled.) To hold. The injunction "Hold

on!"—wait, stop, has probably been taken over from this source into shore speech. See hold the **slack.**

Hole in the beach, to find a. A fisherman's term for reaching a sheltered anchorage in bad weather.

Holiday. A place missed in painting about decks, and used in the same sense alongshore. "Don't you leave no holidays on that fence."

Holy sailor, call the watch! An exclamation of surprise or consternation. (Probably a story went with this once.)

Home. In sea language, home is toward the ship, or inboard. The anchor comes home when it drags; sails are sheeted home. Hence, to bring home to a person some fact that he ought to face, a phrase very common in shore speech. See **Sheet** it home; also **Tumble home.**

Homeward-bound stitches. A term used in reproof to small girls along the coast, when they take long, irregular stitches in their eagerness to finish their "stints" and be off to play. It was originally a sail-maker's phrase, meaning a patched-up job intended only to last until the next port was reached.

Hooker. An old unwieldy vessel; from the Dutch hoeker, a fishing-boat. Often used in affectionate disparagement of perfectly good ships. Sometimes applied to an old prostitute ashore. A drink is also a hooker.

Hook, line and sinker. To take or swallow something hook, line and sinker—i.e., to be gullible—comes from a fisherman's term to describe a fish so hungry that it fairly hooked itself. To hook has a special application to women in search of a husband: "She hooked him good and solid."

Hóoraw's nest. The tangle of ropes on deck when a heavy sea has washed the coils from the pins. Alongshore, it characterizes a condition of extreme disorder: "This room looks as if it would ride out—it's a regular hóoraw's nest."

Horizon. The line which apparently divides the sea and sky. On the horizon, over the horizon, below the horizon, are phrases frequently used in a figurative sense by landsmen.

Horn, the. Cape Horn; to the British sailor, "Cape Stiff."

Horn. See **Blow your horn, Foghorn.**

Hornpipe. An old-time sailor's dance, still taught in dancing schools. To feel like dancing a hornpipe has about the same meaning in shore speech as dancing in the streets.

Horn spoon, by the great! A mild oath which used to be common at sea and alongshore, and which is occasionally used by landsmen. Its significance has been lost, but it probably originated in some tale told on shipboard.

Horse, salt. The salt beef which, in the old days, was the sailor's staple diet. His grace before meat, when the beef kids came in from the galley, ran as follows:

> Old horse, old horse, how came you here?
> —From Saccarap' to Portland Pier
> I carted stone this many a year,
> Until, worn out with sore abuse,
> I'm salted down for sailor's use.
> The sailors they do me despise;
> They turn me over and damn my eyes,
> Cut off my meat and pick my bones,
> And heave the rest to Davy Jones.

The term is sometimes applied alongshore to corned beef.

Horse Latitudes, the. That part of the North Atlantic lying north of the northeast trades; a region of calms and variable winds, where cargoes of livestock sometimes had to be thrown overboard for lack of water.

Horse market. Tide rips, where the meeting tides or currents toss up short, choppy waves.

Hug-the-shore. A timid sailor, or a cautious person. See **Appletree-er.**

Hulk. The body of a ship, used for storage after removal of masts and spars. The word is used figuratively ashore, "a great hulk of a man," or, by extension, "hulking." A sea song popular ashore many years back began "Here, a sheer hulk, lies poor Tom Bowling." That was the shore writer's error: it should have read "shear-hulk," one equipped with a derrick or shears; but if he had grasped the sense correctly, it would have spoiled his meter!

Hull. To inflict damage upon a ship below the water line. See **Hit between wind and water.**

Hull down. Said of a ship so distant that only the masts can be seen above the horizon. Occasionally used figuratively in literature.

Humbug and derricks. See **Windjammer.**

Hurrah's nest. See **Hóoraw's nest.**

Hurricane. A cyclical storm in the western Atlantic. Chase says it is "a word of Caribbean origin that sailors carried to all the tongues of Western Europe."

Hurricanes are most prevalent in late summer and early fall, as stated in the old rhyme:

> June, too soon.
> July, stand by.
> August, look out you must.
> September, remember.
> October, all over.

Huzza. See **Hoist.**

I

Iceberg. A floating mass of ice broken away from a glacier, and constituting a prime hazard at sea. In shore speech, it means a frigid person—usually with "a regular."

In. (1) Of sails, furled and stowed, an abbreviation of taken in. "In the **to'gannels'ls,**" for example, is the order to furl them.

(2) The seamen's use of in and other prepositions in regard to his own service aboard ships is precise and rather intricate. By preference, he goes, or ships, in rather than on a vessel. When he uses the latter word, the phrase board of is either added or strongly implied. Thus, in order of preference:

> I went out boy in the *Carrie C.* for a trip to the
> sealing grounds.
> I shipped as boy aboard of the *Carrie C.*, etc.
> I shipped as boy on board of the *Carrie C.*, etc.

With board of implied but not spoken:

> When I was on the *Carrie C.*, us boys swiped some
> doughnuts from the cook one night, and—

When an officer's rating is mentioned, of is sometimes substituted:

> I shipped next voyage as second mate of the *Vigilant*.

But, with a slight alteration in the wording, we are back at in again:

> Then I went second mate for a voyage with Cap'n
> Gould in the *Vigilant*.

(Don't ask me why—all I can say is, it takes more than a voyage to learn about prepositions!)

Inboard. Over the bulwarks and in upon the deck. Used in the same sense alongshore—of a wagon, for example. See **Board.**

Indraught. The pull of the tide setting to the land.

Irish hurricane. See **Calm.**

Irish pennant. A loose, dangling end of rope. Alongshore, a rag, or anything frayed out and fluttering untidily, is likely to be called by this name. Also called Irish pendant.

Irons. The sea name for handcuffs. "I'll have to put you in irons for that" is a jocular threat alongshore. See **Work one's old iron up.**

When a ship misstayed (see in **stays**) she was said to be "in irons." Irons is also the whaleman's general term for the tools of his trade.

J

Jack, Jack Tar. A generic shore name for men before the mast, but seldom used by seamen themselves. It appears in a few shore phrases, such as "every man Jack," meaning every single one, all hands; and "Jack ashore," or a person in unfamiliar surroundings, one who does not know how to behave. See **Hand** (1), **Mariner, Quarter-deck, Sailor, Seaman.**

Jacksoned. In **irons,** q.v.; or figuratively, in difficulties; from the phrase "hard up (or jammed) in a clinch, like Jackson," the origin of which is not known. Sometimes he was further hampered by having "no knife to cut the seizing." There are other and less seemly phrases about Jackson.

Jetsam. Articles jettisoned from a ship. Common in shore speech, combined with **flotsam,** although without reference to its precise meaning in admiralty law.

Jettison. In admiralty law, to throw overboard cargo to save the ship; used figuratively in literature.

Jib. A triangular sail carried at the bow. A phrase in common use ashore is "the cut of one's jib," i.e., one's general appearance and get-up (usually preceded by "I don't like").

Alongshore, "Up jib!" is a piece of humorous advice to one who seems to be getting into difficulties.

Jibber-the-kibber. See **Mooncurser.**

Jibe (Dutch). To swing the boom of a fore-and-aft sail from one side of the deck to the other, in order to meet a shift in a following wind. A nautical term for a nautical operation, but probably the origin of the term "jibbing" on the part of a horse.

Jig, Jigger. A watch tackle, or extra set of pulleys to be attached where more pulling-power is needed. Another name for it is "handy Billy." When the blocks come together, "the jig is up"—no more can be done.

The phrase has passed into shore speech with no recollection of its original meaning. "I'll be jiggered" may come from the same source or it may refer to infestation by the tropical insects called jiggers (chigoes).

Jolly Roger. The skull-and-crossbones ensign flown by pirates of old. Appears figuratively in shore speech.

Jonah, a. A person whose presence on a ship is believed to bring bad luck. The term is common in the same sense on shore.

Joss-pidgin. Church forms and ceremonies—this is not a gibe at religion as such. Pidgin English, brought home by sailors.

Jumper. Originally a sailor's short jacket, though the name may be connected, through the French jupon, skirt, with the Arabian jubbah. Ashore, it means a woman's sleeveless overdress.

Junk. Worn-out rope and ship's fittings. It was the mate's perquisite to sell it to the "junkman." Both terms have passed over completely into shore speech.

Jury-mast, Jury-rig. Temporary spars and rigging contrived at sea after an accident. Alongshore, the term is applied to any home-made repairs. For a figurative use by a landsman, see page 3.

Just saw it going over the bow schooner-rigged. A humorous reply when asked the whereabouts of a missing article.

Kedge (Pronounced cage). A small anchor; see **Anchor** to windward.

Kedgeree. A dish of rice and fish flavored with **curry;** popular at sea, and now eaten ashore—see any cookbook. It probably received its name from a village on the Hoogly River below Calcutta, mentioned by Rudyard Kipling:

> That night, when through the mooring-chains
> The wide-eyed corpse rolled free
> To blunder down by Garden Reach
> And rot at Kedgeree—

Keel. The long timber at the bottom of the hull which supports the ribs. It appears in several shore phrases: to lay the keel, that is, to outline a plan or found an enterprise; to keel over, meaning to fall in a faint; on even keel, meaning steady or in equilibrium; plenty of water under the keel, i.e., in no need of special precautions.

Keelhauling. A savage Naval punishment in the old days, consisting in passing a rope under the ship's bottom and hauling the offender under the keel from one side to the other. Few survived such an ordeal. Alongshore, it is still a term of indignant reproach to say "He ought to be keelhauled!"

Keelhook. See **Killick.**

Keelson. See **Truck.**

Keep. To steer. Keep her up is an order to steer nearer the wind; keep her off or away means the reverse. The term survives in the shore admonition to keep up. See **Bear** up.

Kettle and coffee mill. See **Windjammer.**

Key to the starboard watch. Same as galley-downhaul. See **Galley** (2).

Kick the bucket. See **Bucket.**

Killick. A stone used as an anchor by small boats. Used

alongshore in the phrase "up killick," meaning to start off in a hurry.

Kink (Dutch). Originally a sailor's word for a twist in a line. It is now common in shore speech both literally and figuratively— one may have a kink in his back, or a kink in his mind.

Kittle o' fish. See **Fish.**

Knock down. The sailor has invented many picturesque variations on the theme of fisticuffs. To "knock down and drag out" expresses the tactics of the **bucko** mate in turning out the **watch,** and has passed into shore speech as the name for a free-for-all fight.

To "knock one galley-west" probably comes via some forgotten tale from **galley** (1) or (2). The origin of "to knock seven bells out of one" has also been forgotten; but to "knock the tar out of one" is a euphemism. All these phrases are used in varying degrees by landsmen.

Knock off. To go off duty, cease working; the nautical opposite of **turn** to. The phrase is in common use ashore, but is probably of sea origin. "To knock off work and carry deals" means to be transferred from a hard job to one no easier.

Knot. The nautical **mile,** when reckoned in terms of miles per hour; a measurement of speed. Shore people sometimes treat it as though it were a measurement of distance. It received its name from the knots, or division marks, in the log line. It figures alongshore in the phrase, at the rate of knots—i.e., very fast indeed.

Knots. See **Bowline, Bends, Granny, Hitch** (2), **Laniard knot, Reef knot, Sheepshank.**

Kyár. Pronunciation of coir, the cocoanut fiber from which some heavy rope is made.

Laden. Of a ship, freighted; having **cargo** on board (obsolete). Ashore, it means oppressed, weighed down, as with grief, care. (Literary.)

Lading, bill of. A term in general use in shore transportation, coming from the old name for a cargo list.

Lammy. A heavy woolen smock, worn at sea and along the coast.

Land. (1) The seaman's observation of the land from the sea has contributed two common phrases to shore speech, "see how the land lies" and "get the lay of the land," both signifying to secure advance information; size up a situation. "So that's how the land lies!" expresses sudden enlightenment. A landlocked harbor is one nearly surrounded by land.

(2) To disembark from a ship. In shore speech, it means to arrive, to find oneself somewhere. "If you keep on this way, you'll land in jail." A seaplane "lands" on the water. The landing of a flight of stairs is from this source.

(As a transitive verb, to land—a contract, a job, or whatever—probably comes from shore fishing.)

Landfall. The first land sighted from sea. The word is sometimes used with a literary flavor to mean arrival at the place aimed for.

Landlubber. The sailor's name for a landsman at sea, particularly an awkward or a seasick one. See **Lubberland.**

Landmark. A prominent object on the coast, recognizable from sea at a considerable distance. In shore speech, it designates a historical event, date or object—even occasionally a person.

Landshark. This originally meant a dealer or boarding-house keeper who swindled sailors, as in the old whaling song:

They send you to New Bedford, that famous
    whaling port

And give you to some landsharks to board and fit
    you out.
They send you to a boarding-house, there for a time
    to dwell—
The thieves they there are thicker than the other side
    of hell.

The term is now in general use ashore to mean money lender,
usurer.

Laniard knot, tough as a. A laniard knot is a heavy knot of
tarred rope securing the stays. The meaning of the phrase is
obvious.

Larboard. See **Port** (2).

Large. Chase defines this word as originally the French nautical term largue, which means having the sheets eased off;
confused in pronunciation with the English word large. See
Wind (2).

Larrup. To thrash. According to Russell, larrup is from an
obsolete nautical word, "lee-rope," but Weekley derives it from
leather or lather, plus wallop, and believes it not to be of sea
origin.

Lash. To tie securely by means of a lashing. Both words are
used in the same sense alongshore, where trunks and boxes
are always lashed, never tied up. Lash is definitely a sea word
when used in the foregoing sense, (though the "whiplash" is not
related).

Late on the tide. See **Tide.**

Latitude, to be out of one's. Of a shipmaster, to have made
an error in navigation. See **Reckoning.** Ashore, the phrase is
used to mean out of one's depth, or not able to "keep one's end
up."

Launch. To set a newly-built ship afloat. This has been a
nautical word since about 1400. Ashore, it means to organize
and promote (a business, movement, campaign or project). By
extension (shore use only), "to launch out upon."

Lay. (1) A share in a fishing or whaling voyage. In this sense,
it is used alongshore in connection with raffles, etc.

(2) Of a rope, the direction in which the strands are twisted, and used in the same sense alongshore. See **Twice-laid.**

(3) An order to the crew to proceed—aloft, forward, aft, "lay out" on a yard, etc. It is occasionally used alongshore in the same sense.

See lay **a course, the keel, lines,** of the **land.**

Lay off. (1) To steer away from the wind. This may be the origin of the shore slang "Lay off!" i.e., quit.

(2) To discharge the crew. In shore speech, it means either to discharge a person, or force him to take a vacation at his own expense.

Lay to. To stop the ship's progress through the water by backing the sails, for the purpose of speaking another ship, waiting for a pilot, daylight, the tide, etc. It usually connotes light weather, but may also be synonymous with **heave to,** q.v. The phrase is well enough known ashore to be the basis of an old music-hall gag about eggs—the ship "laid two."

Lay up. Of a vessel, to take her out of active service, in order to await a charter or to make repairs. Ashore, to be laid up, (laid up for repairs) means sick in bed.

Lead, the. Plummets—the light hand lead and the heavier deep sea (dipsy) which the seaman uses to **sound** the depth of water under the ship. See also **Fathom, Plumb.** On steamships various sounding machines have supplanted them.

The leads have a cup or hollow at the lower end filled with tallow or soap, in order to bring up specimens from the bottom. From the appearance of what comes up on the lead, coasting captains are supposed to be able to make uncannily close guesses as to their position. One mate, skeptical of the "old man's" skill, dipped the end of the lead in the earth that clung to some potatoes, and offered it for inspection. The skipper smelled and tasted it, then sung out "By the Great Horn Spoon, mister, Nantucket's sunk and we're spang over Marm Hackett's garden." Or at least, so they say.

Leading light, Leading wind. See **Light, Wind** (2).

Leak, to spring a, To take in water through the hull. Alongshore, this is said of any container that gets a hole in it, or of a person who urinates involuntarily.

Leave in the lurch. See **Lurch.**

Lee. The opposite of **weather**—the side away from the wind, just as the cross-word puzzles have it. A blanket lee is an especially snug berth. Under the lee of, to mean sheltered by, is common in shore speech. The word was originally lew, hence the pronunciation of leeward, q.v.

Lee set. A low bank of clouds to leeward.

Lee shore. A coast upon which the wind is driving the ship; sometimes used with a literary flavor to describe some danger difficult to avoid.

Leeward (Pronounced loo'ard). The opposite of **windward;** intensified to **dead** to leeward. Alongshore, "to get to loo'ard" means to go in debt or otherwise to lose ground. For leeward ebb, leeward flood, see **Tide.**

Leeway. Rate of drift down the wind, or distance between the ship and some object to leeward. It is used commonly in shore speech to mean scope, room to spare; by extension, to make up leeway signifies to regain an advantage lost.

Leg. The distance covered by a vessel on a single **tack,** when **beating.** Same as **board** or **hitch.** "Leg and leg" means, in shore speech, evenly matched. See also **Show** a leg.

Lend a hand. See **Hand.**

Let her fly, ride. See **Fly, Ride.**

Letters of marque. The official commission of a **privateer** was called a "letter of marque and reprisal." A vessel sailing under such a commission was also called a "letter of marque." The term meets with occasional literary use—as the title of a book of travels or essays, for example.

Libel. In admiralty law, to place a lien upon vessel or cargo. Alongshore, to attach property in satisfaction of debt; garnish wages. (No connection with the shore use of the term, to mean written slander.)

Life line. A rope strung along the deck in heavy weather, as a handhold to prevent being washed overboard. Also a life-saver's line from shore to ship. It appears in shore speech in many figurative uses.

Life preserver, Lifesaver. A float thrown to a person who falls overboard. In gangster slang, life-preserver means a blackjack. Lifesaver is used ashore in all sorts of figurative ways.

"A perfect lifesaver!" (Also as the proprietary name of a small candy shaped like a life belt.)

Lifts and braces. See **Set up.**

Ligan. See **Flotsam.**

Light. A lighthouse. Probably the origin of to see a light, i.e., to comprehend; also leading light, a prominent person—phrases common in shore speech.

Lighter. A barge used to transfer cargo between a vessel and the shore.

Nantucket has a rather incomprehensible term meaning belated—astern the lighter. "You're astern the lighter a'ready with all your garden chores."

Light out. Go, or carry—used in such orders as "Light out (to windward, leeward)" in furling sail, or "Light it along," in carrying unbent sails on the deck. Apparently this use of light is strictly nautical. It is probably the origin of the colloquial phrase "Light out!" i.e., get out of here fast!

Lights. Night signals on shipboard. Side lights or running-lights are red and green lights shown to port and starboard respectively. The lookout on the forecastle head, as he repeats the helmsman's strokes on the bell (see Time), looks at the side lights and sings out to the officer on watch, "Lights burning bright, sir!"

The Rules of the Road for night sailing are:

When both lights you see ahead,
Port your helm and show your red.

Green to green or red to red,
Perfect safety—go ahead.

If to your starboard red appear
It is your duty to keep clear.

But when upon your port is seen
A vessel's starboard light of green,

There's nothing much for you to do,
For green to port keeps clear of you.

When the ship is at anchor, a white riding-light is shown on the foremast, and there are other signals to indicate that she is being towed, is unmanageable, etc. See **Distress.**

Limejuicer, Limey. An Englishman; from the anti-scorbutic formerly served out on British ships. The term is frequently used by landsmen, especially by our expeditionary forces.

Line, the. The equator. The line storm is a heavy gale expected on the Atlantic coast about the time of the autumnal equinox, when the sun crosses the equator.

Line. See **Rope.** For all except a few definitely named ropes, the word line is preferred both on shipboard and alongshore.

When the whaleman has his harpoon in a whale, he is said to have "got a line on it." The phrase has passed into common speech ashore, meaning to begin to grasp the import; or to have taken the first steps. (But "to line up" is not nautical.) See **High line, Life line.**

Lines. The design of a ship; used alongshore in describing a person's figure. See **Model.** The master builder "laid down the lines" of a vessel before starting to build her, and the phrase has passed into shore speech to mean drawing up plans of any sort. On these lines, along the lines of, and hard lines apparently come from this nautical source. (To line out—i.e., sketch a plan—probably comes from the same source, but may refer to finding honey by tracking wild bees. To line out in playing baseball, however, has no relation to either.)

Lingo. A sailor's term, meaning jargon. It was probably brought home from the Mediterranean, where "lingua franca," a mixture of many tongues, is generally used. The word is now incorporated into shore speech, generally with a disparaging implication.

List. A list, meaning an inclination, was formerly a shore word, but about the only trace of it left in shore speech is the archaic "if you list." It is now completely nautical. To list is to tip to starboard or port, owing to improper stowage or shifting of cargo or ballast. Alongshore, it is used to characterize any deviation from the perpendicular. "He walks with a list" means that he limps; but to say, "he's got quite a list on" means that he is intoxicated.

Liverpool button, a. A toggle of wood used to replace a missing button on a seaman's clothing. The term probably refers to the "tinkers and tailors and sojers and all" who, according to the old song, shipped as sailors on the Western Ocean packets, and whose clothing was not of the best. The term is occasionally, though rarely, applied to some makeshift alongshore.

Liverpool, see you in. "See you in Liverpool" is an offhand farewell greeting (still used by pilots in leaving outbound vessels) which dates from the day of the Western Ocean packet ships.

Living gale. See **Gale.**

Lobscouse. The sailor's term for a hash of meat, onions and hard bread, pounded up. Smyth says it was originally "Lapp's course." It is applied jokingly alongshore to hash of any kind. "Are you going to have lobscouse for supper?"

Lobster, red as a boiled (Pronounced b'iled). From lobster fishing, and now common in the speech of persons who never saw a lobster in the shell.

Locker. A compartment for stowage aboard ship, as chain locker, paint locker, sail locker. Use of the term at sea probably antedated its present shore use, to mean locked compartments for holding personal belongings in shower-rooms, schools, industrial establishments, and so on. See **Boatswain, Davy Jones' Locker, Shot.**

Lodeman. See **Pilot.**

Lodestar, the. Polaris, the North Star, used by mariners of old as an aid to navigation. It was supposed to attract and hold the compass needle to due north. Its use ashore is purely literary, "the lodestar of one's affections," etc.

Lodestone. Magnetite with magnetic properties, used as the forerunner of the mariner's compass by the Norsemen, Arabians, Chinese and other early nautical races. Floated on a piece of wood in a basin of water, it pointed to the north.

Log. (1) An instrument for measuring the speed of the ship. Nowadays it is a mechanical device, on the principle of the speedometer; but the old-time log consisted of a wooden float, the "log chip," attached to a marked line which was wound on a spinning "reel." The log was thrown overboard, and at the end of a definite number of seconds, told by a sandglass running out, it was ascertained from the marks on the log line how far the ship had gone. From this her rate of speed in knots, how much she was "logging," was calculated. See **Knot.**

When the ship was going fast, the reel gave off a humming note. Hence has come into shore speech "to reel off" a statement, and "right off the reel," i.e., without stopping.

(2) The word log also means the ship's official diary of weather and events, usually kept by the chief officer. (It got this name from the "log board" or slate, on which the readings of the log (1) were entered. In Falconer's day, it was called the ship's "journal," a much better title.)

To be logged is to have one's derelictions entered in the log-book. Alongshore, "I'll log ye for that" is a threat to report one. One mate, who was unable to write up his log because of being "three sheets in the wind," found the next day that the captain had taken over this duty, and had logged him as being drunk. He said nothing, but at the end of the next day's write-up he added "Captain sober today."

Every entry in the log ends with the conventional phrase "So ends this day." In one old-time log the mate entered among other details, "Captain sick. Mixed him up a dose of physic." The next day appeared the laconic entry, "Captain dead. Packed him up in salt. So ends this day."

The log, in this second meaning, appears in shore speech as the name of a process in making industrial time studies.

Loggerheads, at. Loggerhead has two different meanings, either one of which may be the parent of the shore phrase "at loggerheads," i.e., quarrelling or in disagreement. A loggerhead was a post in a whaleboat to which the harpoon line was made fast when the whale was "ironed"; it was also an iron rod termi-nating in a ball which was heated and used to stir flip and other drinks.

London River. The Thames is known by no other name than this to sailors.

Long-sparred. See **Spar.**

Look alive! A term, probably of sea origin, used both at sea and ashore to mean "Step lively" or "Get a move on."

Look up. Of a vessel, to head closer to the wind. The shore phrase, "things are looking up," i.e., improving, may come from this source.

Lookout. A member of the crew stationed on the forecastle head to keep a watch for possible danger. The shore phrases "to keep a bright, or good, lookout" and "Look out!" come from this source. See **Squalls.**

Loom. The shape of the land, another ship, etc., seen through fog. Used in the same sense alongshore, both as a noun and verb. "I saw the loom of the land." "The house loomed up in the fog."

Loose. See **Adrift, Cut loose from.**

Loose ends, at. A shore phrase, meaning without objective, aimless, or disordered. It probably refers to a rope cast off the pin and dangling idly.

Loss, a total. In admiralty law, a shipwreck with no possibility of saving the hull. When this exact combination of words is used to describe disaster ashore, the probability is that it comes from sea language.

Low bridge. A warning to duck the head, often heard on landsmen's lips. It comes from canal boating—if that can be called a branch of seafaring!

Lubberland. An imaginary country of sailor mythology, similar to **Fiddler's Green,** q.v. A village near Portsmouth, N.H., formerly went by this name. "Lubber," though it figures largely in the seaman's vocabulary, does not seem to have been originated by him. But see **Landlubber.**

Luff. To steer closer to the wind. Occasionally used alongshore to mean "Pull over to the side of the road."

Lugs, to put on. Shore slang (obsolete) meaning to put on airs. It may be connected through some forgotten story with the lug sails used on some small vessels. (The other meaning of lugs—ears—doesn't seem to fit so well.)

Lunar. To take a lunar, in navigation, is to take a **sight** on the moon at night instead of the usual noon observation of the sun's altitude. Alongshore, to take a lunar means to take a turn around the yard at night and size up the weather before turning in.

Lurch. A heavy roll to windward or leeward—originally a seaman's word, "lee-larch." It is common in shore speech—a lurching gait, etc. (To leave in the lurch, however, is derived from an obsolete game similar to backgammon, and not from the sea.)

Mackerel sky. Mottled cirrus clouds foretelling a change of weather.

> Mackerel skies and mare's tails
> Make tall ships carry low sails.

Maiden voyage. See **Voyage.**
Main, the. An islander's term for the mainland. "I'm going over to the main for the week end."
"The main" as a name for the ocean is purely literary.
Main brace. See **Splice.**
Mainmast. The second mast from the bow. Deaf as the mainmast is a 'longshore phrase meaning the same as deaf as a post to the landsman.
Mainsail haul. "Mainsail haul!" is the order for the most important operation in tacking ship. Alongshore, it has come to mean a great stroke of luck or profit. "I'm going to make a mains'l haul this hand."
Mainstay. The principal rope in a ship's rigging, holding the mainmast in position. Ashore, it means the chief support or reliance.
Make. (1) Use of this verb, sometimes intransitively, to mean grow, increase, seems to be peculiar to the sea and the coast. The tide makes—i.e., comes in; a storm cloud, or the sea, makes up—i.e., gets rougher or more threatening. A point of land makes out from the coast; a bay makes in.

The helmsman is ordered to make (i.e., strike) eight bells or whatever the time. The vessel makes (leaks) so-and-so many inches of water. In signallizing another ship, she "makes her number" with code flags. See **Signals.** In getting under way, she makes sail. A related shore use is to make ready, or prepare.

(2) To espy; the sailor makes a light or the land, or he may

make them out with difficulty. See **Pick up.** A phrase in coastal dialect, to make out ("I can just about make out" to accomplish something) comes from this meaning of the word.

(3) To reach, attain. The vessel makes the port she has previously made for (and then makes fast to the dock). See **Hitch.** In sailing, she makes **headway** or possibly **leeway.** An officer acknowledges a report from an inferior on the course, the time in navigation, etc. with a curt "Make it so!"

When a shore person says he must make for home, or boasts "I made it!" after catching a bus, he is using sea terms. See **Weather** (2).

Makee Look-see. To reconnoiter, search for. Pidgin English brought from China by sailors.

Man. (1) A common sailor. See **Hand.** The steersman is the "man at the wheel," a phrase sometimes used ashore to designate a director or person in command.

(2) To take station as ordered, at the ropes, boats, guns, or whatever. Manning the yards—stationing men along them standing at attention—is an especial courtesy to dignitaries. To man is sometimes used, both literally and figuratively, in shore speech.

Man Jack. See **Jack.**

Man-man. Wait a bit; go slow. Pidgin English; not one of the phrases in common use alongshore, but understood if used by a sailor home from China.

Man overboard! A dread sound at sea; but the sailor has his joke about it as about all other hazards. One Barnabas, the mate of a small coaster, was knocked overboard by the jibing of the boom, and the hand at the wheel tried to give the alarm as the captain rushed up from below. "B-B-B," he said. "Sing it, you fool!" yelled the skipper; and the steersman chanted, "Overboard is Barnabas, and half-a-mile astern of us!"

The phrase is used jokingly in shore speech when someone falls off, or something is dropped.

Man-o'-war. A warship. Man-o'-war fashion means very smartly; according to a definite ritual. It is used figuratively in this sense alongshore. The **frigate** bird was formerly called the man-o'-war bird.

Manavelins. Small stores on board ship; the sailors' contemptuous term for fancy, fixed-up victuals, and used in the same sense alongshore.

Maneuver, manoever. To handle a vessel under sail. Honors are about even whether this began as a military or naval word; it was borrowed as one or the other from the French in the 1700's. (Weekley.) In shore speech, it has taken on a slightly sinister cast—to plot or intrigue.

Manhandle. This word originally meant to move cargo, etc., by sheer manpower, without use of mechanical aids. Figuratively, it means to handle roughly; to administer physical abuse. It is probably not, however, of nautical origin.

Manifest. An exact list, for customs purposes, of items carried in a ship's cargo. See bill of **lading.** It is sometimes applied alongshore to similar lists—for example, the inventory of a store.

Manila. A light-colored rope of fine quality, made of fiber originally grown in the Philippines, and in common use under this trade name, ashore as well as at sea.

Mares' tails. Feather-like cirrus clouds foretelling bad weather. See **Mackerel sky.**

Marine. (1) Having to do with ships or the sea. See **Maritime, Naval, Nautical.**

(2) As a noun, "marine" may mean the shipping of a country (naval marine, **merchant marine**).

Or (3), it may mean a soldier stationed on a warship. These men knew nothing of seamanship and were easily hoodwinked by the sailors, hence, "tell it to the marines" ("I'm too smart to be taken in" implied), a phrase common now in shore speech. Sailors used to call an empty bottle a "dead marine." See **Soldier.**

Mariner. This word originally meant a person skilled in the craft of the sea, as opposed to the mere seaman (and still preserved in this sense, in the term **master** mariner). Other uses of the word in shore language are entirely poetical, except when, as sometimes happens, shore people confuse mariner with marine. See **Merchant marine.**

Maritime. About the same as **marine,** but also meaning coastal, as in the Maritime Provinces of Canada. See also **Naval, Nautical.**

Marks. Symbols painted on vessel's hull to show loading depth. When loaded, she is "down to her marks." Alongshore, this might be said of a hayrack or some other vehicle.

Marlinespike. An implement used in working with ropes, etc., which appears in several 'longshore phrases. "It would hold up a marlinespike point down," is said of strong coffee, thick pea soup, etc. Some physical impossibility may be likened to "sitting on the point of a marlinespike." See **Sailor, Spike.**

Maroon. To set a person ashore and desert him—a custom of the buccaneers.

Marooned is fairly common in shore speech to signify the plight of a person isolated or stranded in unfamiliar or distasteful surroundings. In the West Indies, by some perversion of ideas, a maroon has come to mean a picnic, particularly a sailing-party.

Martingale. A small boom extending downward from the bowsprit. Ashore, part of a horse's harness. There is no way of knowing which came first; one or the other may have come from the Spanish almartaga, a halter. There is a suggestion also that the name may have come from the town of Martigues in France, in reference to a peculiar kind of breeches worn by the inhabitants.

Maskee. All right; no matter. Pidgin English—not in common use.

Mast. This word appears in one shore phrase, "tied to the mast," i.e., helpless, at the mercy of chance—as a sailor lashed to a spar and drifting helplessly after shipwreck. See **Before the mast, Colors.**

Mast paint. Pea soup—from the yellowish paint sometimes used on masts.

Master. An abbreviation of the rating master mariner, or the captain of a merchant vessel. Formerly a title of address, but no longer so used. See **Forms of address.** See also master **hand.**

Masthead high. A coastal phrase meaning very far up.

Mate. "The mate" (originally the "master's mate") means the first or chief mate of a vessel. It is sometimes applied jocularly alongshore to a wife—though in the nautical, not the dictionary sense. See **Forms of address.**

Mattalow. An old-timer; a thorough seaman. French mate-lot, a term brought to the coast by seamen. Often preceded by "a regular old." See old **Salt, Shellback.**

Mayday. See **S O S.**

Medicine chest. The ship's medicine chest, together with the "doctor's book" that accompanied it, comprised the ship-master's equipment for treating illness at sea. Sometimes the bottles were simply numbered, the book giving directions for dosages according to the symptoms. One captain, finding a dose of No. 10 indicated but the bottle empty, is said to have handed the sick sailor a double dose of No. 5. It was a healthy life, and he probably recovered.

A cabinet containing household remedies is always "the medicine chest" in 'longshore homes.

Merchant marine, the. Merchantmen, that is, vessels engaged in cargo carrying in distinction from naval vessels. The term is understood by some shore people, but is apt to be confused with the Marine Corps, a branch of naval personnel.

Mermaid. A fabled creature, half woman and half fish, which appears in the folklore of all lands, and which was firmly believed in by sailors at least until the 19th century. The mermaid legend has been ascribed by some to observations by early explorers of the manatee, a small cetacean found in Caribbean waters, which has the curious habit of rearing itself on end part-way out of the water. The probability is, however, that the legend is as old as that of the **siren,** q.v.

Merry men. The name given in some parts of the world to **tide rips,** q.v.

Mess. In shore speech, a joint ménage for eating purposes, or to eat together habitually (said of nonfamily, cooperative groups). This was originally a sea term, messmates being those who ate together from mess-kids, or wooden serving tubs. To lose the number of one's mess means, figuratively, to die. It is occasionally found in shore speech or literature. See **Everybody's mess.**

Middle Passage, the. The Atlantic Ocean between Africa and the West Indies. This was the middle trip of the slaver's triangular voyage.

Middy. Abbreviation of midshipman, a boy serving apprenticeship to become an officer in the navy or merchant marine. The girl's garment called a middy blouse is modeled on his uniform.

Midships. See **Amidships.**

Mile. The sea mile is greater than the land mile. It is based on the length of a minute of latitude or longitude, a measurement that varies at different parts of the earth's surface. Different countries assign it varying lengths, all, however, in the neighborhood of 6000 feet—a multiple of the **fathom** and **cable's length,** q.v. The U. S. Coast Survey adopts a value of 6080.27 feet, the British Hydrographic Office, 6080 feet. See **Knot.**

Misstay. See **Irons, Stays.**

Mister. See **Forms of address.**

Model. The plans and specifications for a ship's hull (or a miniature facsimile of the ship). See **Lines.** Alongshore, the word is used in reference to a person's figure. "She gets around pretty lively for a woman of her model."

Moderate. Of wind and weather, to subside, become more pleasant. The presiding officer at a New England town meeting is called the moderator.

Monkey-pea-jacket, a. A facetious term alongshore for any outer wrap. From monkey jacket and pea jacket, garments figuring in the seaman's wardrobe.

Monsoon. A periodic wind in the Indian Ocean and China Sea, blowing six months in one direction and six months in the other. It is frequently confused by landspeople with cyclones and other circular storms.

Montyreevo (Spanish monter arriba, go up quickly). Shouted at men going aloft, when they lagged, and occasionally heard alongshore.

Mooncurser. A South-of-England name for a wrecker, who enticed vessels ashore by tying a lantern to a horse's bridle and hobbling one leg, so that the animal's stumbling simulated the motion of a vessel—a practice known as "jibber-the-kibber." This was not, of course, feasible on fair moonlit nights, hence mooncurser.

American sailors apply the name chiefly to the inhabitants of the "back shore" of Cape Cod, though they do not accuse these people of jibbering the kibber—the grim old Cape requires no such assistance from the hand of man. (See Henry C. Kittredge's *Mooncussers of Cape Cod.*)

The story is told of one of these towns that on a stormy Sunday morning while church service was in progress, a man ran up to the church door and shouted "Wreck ashore!" The minister called out authoritatively, "Keep your seats until I have pronounced the benediction," in the meantime making his way down the aisle as rapidly as his dignity permitted. Pausing at the door, he gave the blessing, adding, "And now, my friends, let's all start fair!"

Moor. To secure a vessel so that she cannot swing, by making her fast to buoys or using anchors at bow and stern. Said Publius Syrius in 42 B.C., "It is well to moor your bark with two anchors." Boats are said to be moored when they are fast to a single buoy, however.

The use of moor to mean attach, about other than vessels, is confined to poetry and literature; but "adrift from one's moorings" is occasionally heard in shore speech.

Mosquito fleet. A collection of small craft. The term is seen occasionally in the newspapers.

Mother Carey's Chicken. The sailor's name for the stormy petrel. It has been said to come from the Spanish madre cara, but Weekley says this is "unsupported by evidence." The term was used by Charles Kingsley in *Water Babies.*

Mud hook. The seaman's slang term for the **anchor.**

Mug up, a. A hearty supper or between-meal snack, on fishing vessels. Occasionally heard on shore.

𝒩

Nantucket sleigh-ride. This term refers to a whaleboat fast to a whale, which sometimes runs away and tows the boat at a furious clip for miles.

Nautical. Pertaining to ships, sailors, and navigation. See **Marine, Maritime, Naval.**

Naval. Pertaining to warships rather than to merchant vessels, except in some special uses; e.g., naval stores which consist of tar, pitch, turpentine, and the like. See **Marine, Maritime, Nautical.**

Navigation. Narrowly defined, navigation is the mathematical process of ascertaining a vessel's position by observation of the heavenly bodies. See **Sight.** A good navigator must be able, however, to lay a **course** based upon his observations, and that involves a further skill and knowledge of weather, wind and current. When the historian Gibbon said, "Winds and waves are always on the side of the ablest navigator," he was speaking figuratively; but it can be applied literally as well.

There is a large common area between navigation and seamanship (see under **Seaman**) but they do not exactly coincide. A man might be a good seaman without being able to handle a sextant; or he might be a good navigator and yet be careless about the upkeep of his ship, or unable to "get the most out of her" on a passage.

Alongshore, to navigate is used with several special meanings:

To get about in close quarters, "My kitchen's so small I can hardly navigate around the stove."

To be reasonably sober, "able to navigate."

To carry on affairs. "I'll be able to navigate as soon's I get my insurance money."

The British navvy, or common laborer, was jokingly named "navigator" when he was employed in large numbers digging canals.

Navy blue. A popular color ashore, that gets its name from the naval seaman's uniform.

Navy yard. A place maintained for the repair and refitting of naval vessels. The "Navy Yard District" in many coastal cities is a lively quarter.

Neap, Neaped. See **Tide.**

Newfie. A native of Newfoundland.

News. "Do you hear the news?" was the old formula used after calling the watch—signifying "Are you awake?" and requiring the reply "Aye, aye!" It used to be used jokingly alongshore, with dilatory risers.

Nigh, Near. See **Wind** (2). "No nigh" was an old-time order to the helmsman not to steer the ship closer into the wind.

No bono (Spanish bueno). An emphatic "no good"—a phrase brought home by sailors.

No breakfast, No supper. A derisive term applied to vessels hailing from New Brunswick and Nova Scotia, from the initials N. B. and N. S. appearing on their sterns.

Northing ("th" as in clothing). Progress to the north. A vessel makes northing.

Northward (Pronounced no'th'ard; "th" as in clothing). Toward the north.

Oar(s). Used in many shore phrases:

To **pull** the laboring (i.e., the leeward) oar, meaning to do the heaviest share of the work; to bend to one's oars, or put forth diligent effort; to pull a strong oar with someone, signifying to stand in high favor with him; to lie, or rest on one's oars—to cease effort temporarily; to shove, put, or stick, one's oar in, that is, to intervene without being asked.

Ocean. "The ocean" is not in the sailor's vocabulary. He uses the word only as part of the name of a definite Ocean, or in combination, as in ocean-going vessels.

A very early term for the waters lying west of Europe beyond the Mediterranean was "the ocean-sea." Samuel Eliot Morison called his recent book about the voyages of Columbus *Admiral of the Ocean Sea.* So unfamiliar was this term to shore ears that the title appeared in one magazine as *Admiral of the Open Sea.*

Off and on. A ship tacking toward and away from the land, waiting for daylight or the tide, is said to be standing off and on. Used in shore speech (without knowledge of its origin) off and on means now and then, intermittently.

Offing. Distance from the land, especially a harbor or port. Among landspeople, in the offing means contemplated, in prospect.

Off the wind. See **Wind** (2).

Oil on troubled waters, to pour. In extremely heavy weather, a bag of oil is sometimes hung outboard and allowed to drip so as to form a film on the water and prevent the waves from breaking over the vessel. Use of this phrase by landspeople, meaning to soothe, tranquilize, comes from this sea practice.

Oilskins. Waterproof garments of oiled cloth used at sea and alongshore; slickers. Rope-yarns tied about the wrists and ankles to keep water out are called soul-and-body lashings.

Old Man, the. The sailor's name for the captain, whatever his age—so common as to convey no disrespect when used as a term of reference. It is sometimes applied ashore to the "boss" or head of a concern. See **Forms of address.**

Old Shiver-the-Mizzen. A term of affectionate disrespect at sea and alongshore.

On the wind. See **Wind** (2).

Open-and-shut. Said of alternate spells of clear and cloudy weather. "Open and shet—sign of wet," they say along the coast.

Open up. This phrase is used in connection with landfalls. It can be used as a transitive verb: "When you open up the light, keep her off due west"; or it may be used intransitively: "As the harbor opened up, it was seen to be full of shipping."

In shore speech, to open up means, figuratively, to begin to explore a subject.

Orders. To obey, disobey, orders, and "until further orders" (i.e., indefinitely)—phrases in common use ashore—probably have a nautical rather than a military origin. A sardonic aphorism at sea and alongshore was "Obey orders if you break owners." See **Sailing orders.**

Ordinary seaman. O. S.—one able to **hand,** reef and steer. See **Sailor, Able seaman.**

Outboard. Outside the ship and alongside. A detachable propeller for small boats, with a motor attached, is called an outboard motor. See **Board.**

Out East. The Orient in general.

Out of one's reckoning. See **Reckoning.**

Outward bound. On the passage away from the ship's home port or country of hail. Ashore, it is figuratively used to mean dead, as in the well-known play of the title.

Overboard. Over the ship's side and into the water. In shore speech, to throw, toss, or cast, overboard means to abandon, desert. See **Board, Heave, Man overboard.**

Overhaul. A strictly nautical word, with several meanings:

(1) To let out rope through a tackle; used in this sense alongshore, but not by shore speakers.

(2) To overtake and come up with another vessel. Alongshore, to overtake on the road.

(3) To look through and put in order, as a locker or chest. The shore phrase "a thorough overhauling" comes from this use.

Overalls, or working clothes, are commonly called "overhauls."

Overset. See **Capsize.**

Oversparred. Having masts too tall for the size of the hull. Alongshore, it is said concerning anything top-heavy.

Overwhelm. From an Anglo-Saxon word meaning to bury in heavy seas.

Packet. See **Flash.**

Paddy's hurricane. See **Calm.**

Painter. The rope by which a small boat is made fast. To cut one's painter is to play a practical joke upon one—a phrase occasionally heard on shore lips as well as alongshore.

Palaver. Portuguese palavra, meaning word; used by early Portuguese navigators to describe parleys with the natives in West Africa, where it became incorporated into the native pidgin as palaver, and was later brought home by English seamen. In shore speech, it carries the implication of insincere flattery.

Palm-and-needle. Sailors' tools for sewing sailcloth. The palm is a bone or metal thimble set in a leather hand strap; sail-needles are stout, three-sided affairs with sharp points. Alongshore, it is applied jokingly to sewing implements. "Get out your palm-and-needle, Mother, and set a patch in these pants for me."

Pampéro. A tempest that attacks without warning off the **River** Plate; hence, alongshore, a brief and violent quarrel, a tantrum.

Paper Jack. See under **Dry nurse.**

Papers. A ship's papers are all her documents as to hull and cargo. To take out papers, however, means to procure the official documents necessary to put a ship in commission. It may be the origin of the same phrase used in connection with securing other documents—for example, citizenship papers. See **Clearance papers.**

Parbuckle. To roll a cask or similar object by means of slings. The word was originated by sailors, but it and the process are familiar to brewery workers and others who handle barrels ashore.

Parcel. In the sense of wrapping to prevent chafing, to parcel is a nautical word. It is used in the same sense alongshore.

Part. The seaman's standard equivalent for the landsman's **break,** as applied to ropes, etc. See **Carry away.** For his word as applied to rigid objects, see **Stave.** See also part **company.**

Part Brass-rags. According to Bowen, this phrase, now common in shore slang, and meaning to break up a friendship, comes from the seaman's habit of sharing brass-polishing rags with his best friend aboard.

Passage. Not a word of sea origin, but used to mean any sort of journey. In sea language, it means a vessel's run from port to port. "A pleasant passage and a safe return" is the customary farewell formula among shipping people to those departing on a voyage, and they often use it in connection with land journeys as well.

To work one's passage, a phrase common in shore speech, carries an overtone of being exploited, and expected to do more than one's fair share. "I guess she has to work her passage, all right."

Passenger. Like **passage,** this word did not originate at sea, but as used in land phrases, "Oh, I'm nothing but a passenger; he does all the work," it probably harks back to the sea, as most certainly does the Cape Cod phrase of commendation to a small boy, "That's the little gentleman passenger!"

Passport. This word, used in Hakluyt's *Voyages,* first meant a permission to trade, given in time of war by naval commanders to merchant vessels. In modern shore speech, it means official permission to citizens to travel outside their own country. The English word has been adopted into most European languages.

Passy-ar, a. Spanish pasear, brought home from the Mediterranean or South America by sailors, meaning a walk or stroll. "Guess I'll take a passy-ar and makee look-see."

Patch on a patch and a patch over all. A phrase used to describe much-mended sails and, alongshore, garments.

Pay. To fill the seams of a ship with tar or pitch. Used in the same sense alongshore. See **Hell to pay.**

Pay off. (1) To let the ship fall off from the wind.

(2) To discharge, and pay accumulated wages to the crew. The shore phrase "I'll pay you off for that" (i.e., get even with you)

may be related. Alongshore, to pay off with the topsail sheet means to skip without paying one's debts.

Pay out. To let out the slack of a rope. Used in the same sense alongshore. A person talking bombastically is said to be "just paying it out."

Pea jacket. See **Monkey.**

Peaked Hill Bars (Pronounced pick-ed or picket). A dangerous reef off Cape Cod. (If peaked means sharp-pointed, it is always pronounced as above on the Maine coast. If pronounced peek-ed, it means pale and sickly-looking.)

Pearl diver. A dishwasher on a passenger steamer. The term has been adopted by restaurant workers ashore.

Philadelphia catechism. See **Work old iron.**

Pick up. To succeed in seeing something at a distance. See **Make (2).**

Pieces, to go to. Said of a ship completely demolished by the sea. This may be the origin of the shore phrase, go all to pieces, become emotionally upset.

Pierhead jump, a. A sailor who joins the ship at the last moment before sailing is said to have taken a pierhead jump. Alongshore, the phrase is used to describe any sudden start, unprepared for and unexpected.

Pile Up. The sailor's comparatively recent catchword for running a vessel aground, and first used in that connection, though later applied to automobiles, and to airplanes before the word crash made its appearance.

Pilot. (1) A man familiar with local waters, but stationed on shore, who boards vessels and guides them into and out of port. Before 1500, such a man was called a lodeman, while pilot meant a ship's helmsman.

The man at the controls of an airplane has the official title of pilot. He is not generally, however, a **sky pilot** (q.v.)!

(2) To pilot means to steer or guide a vessel into or out of port. Ashore, the word is used to mean guide or direct. "I'll be glad to pilot you through Chinatown."

Pilot bread or biscuit. Crisp round crackers, currently sold ashore under this trade name.

Pintle. See **Gudgeons.**

Pipe down! Navy slang meaning "Shut up!" Reference is to the boatswain's pipe. It is now common in shore slang.

Pipe one's eye, to. An obsolete phrase meaning to weep. It was probably of nautical origin.

Piracy. In admiralty law, this means robbery on the high seas. In shore speech, it has come to mean aggravated robbery, particularly of ideas. "That's sheer piracy!"

To pirate is a made-up shore word, meaning to infringe a copyright or patent.

Pitchpole. To turn over end for end, as a small boat in the breakers, or a whale in breaching. Alongshore, it means to turn a somersault.

Place for everything. "A place for everything and everything in its place" is the motto of a well-kept ship. The phrase is often heard alongshore as a reproof for disorderliness. See **Shipshape.**

Plain sailing. A landsman's term, meaning clear, uncomplicated (usually with "all"). This is a corruption of the old-time seaman's plane sailing, i.e., by chart instead of by dead **reckoning.** Landspeople have corrupted the term still further, into straight sailing, an expression no seaman ever used. See **plain sail.**

Play fast and loose. See **Fast and loose.**

Plug. Probably a sailor's word originally, for what he used in stopping leaks. It is now firmly incorporated into shore speech. To plug away at something is thought to be named from the sound of oars; but "an old plug" (a horse) from the clop-clop of its hooves.

Plumb. At sea (obsolete), to throw overboard a plummet in order to ascertain the depth of the water. See **Fathom, Lead, Sound.** In shore speech, to plumb the depths—of misery or whatever—undoubtedly comes from this sea use.

When plumb is used as an adjective to mean straight up and down, it is a carpenter's rather than a sailor's term, and this is probably the source of shore uses meaning out-and-out—plumb discouraged, "plumb loco," and so on.

Points of the compass. See **Compass.** The points of the compass, together with their pronunciation by sailors and coastal people, are shown below.

Compass Points	Pronunciation
North	North (sometimes nothe, to rhyme with loathe)
North by east	Nothe by east
North-north-east	Nor'-nuth-east (as in another)
North-east by north	Nothe-east by nothe

Compass Points	Pronunciation
North-east	Nothe-east
North-east by east	Nothe-east by east
East-north-east	East-nothe-east
East by north	East by nothe
East	East
East by south	East by southe (as in the verb to mouth)
East-south-east	East-suth*-east (as in another)
South-east by east	Southe-east by east
South-east	Southe-east
South-east by south	Southe-east by southe
South-south-east	Sou'-suth*-east
South by east	Southe by east
South	South or southe
South by west	Southe by west
South-south-west	Sou'-sou'-west
South-west by south	Sou'-west by southe
South-west	Sou'-west
South-west by west	Sou'-west by west
West-south-west	West-sou'-west
West by south	West by southe
West	West
West by north	West by nothe
West-north-west	West-nor'-west
North-west by west	Nor' west by west
North-west by north	Nor'-west by nothe
North-north-west	Nor'-nor'-west
North by west	Nothe by west

East and west are always pronounced as in shore speech, but north and south are only thus pronounced when they stand alone. When used in combination, they are markedly altered in pronunciation.

---

* Sometimes Southe.

Between each point are half-points and quarter-points, the use of which can be illustrated by a story told about Captain Paul West of Nantucket. Coming on deck one morning, he asked the man at the wheel, "How does she head?" "Nor'-west by west, half west, a little westerly, sir," the man answered. "If you could get another west into that course, I'd give you enough canvas to make a hammock," the captain remarked. Quick as a wink, the sailor answered, "Nor'-west by west, half west, a little westerly, Cap'n West."

Poop downhaul. Same as **galley** downhaul.

Porpoise, fat as a. A common comparison alongshore to this frisky cetacean.

Port. (1) A seaport. The port of destination is that to which a vessel is bound; a port of call is one visited en route. See **Put** in, **Touch at.** A vessel's home port, or port of hail, appears on her stern. All these phrases appear occasionally in shore speech, together with "any port in a storm."

(2) As a direction, port means to the left hand side of the vessel. It was developed in fairly recent times to take the place of larboard, because this word is so easily confused with **starboard.** "Port your hellum!" is a jocose order sometimes given in the coast towns.

Portygee. Pronunciation of Portuguese. A "Portygee man-o'-war," frequently but erroneously called "nautilus," is a small shellfish on the cuttlefish order, found floating in tropical Atlantic waters. It looks like a tiny ship of mother-of-pearl. An earlier name for it was "Salleeman," from the galleys of the pirates of **Morocco.**

Press into service. A shore expression, meaning to utilize, adapt, or co-opt. It stems from the press-gangs of the British Navy in the 18th century, which kidnaped English and even American citizens to serve on warships.

Pride of the morning. A light fog at sunrise, presaging good weather. (This poetic-sounding phrase has another current significance, and its use in mixed company is not advisable.)

Privateer. A privately owned vessel, formerly commissioned (often by **letters of marque**) to take in prize, vessels of the enemy in time of war. Privateering is occasionally used by landspeople to characterize lawless or predatory behavior.

Pronto. Spanish, meaning quickly, at once. It may have been introduced into shore speech, where it is common, from the Mexican border, but as used by coastal people it was doubtless brought home by sailors, who use it freely.

Prow (Latin prora, via French proue). A literary archaism for a ship's bow and still used by landlubbers who know no better. The only "prow" known to the sailor of today is the proa, a Malay sailing-craft.

Pull. (1) A tug upon a rope:

> I thought I heard our chief mate say,
> "Give one more pull and then belay."

Pull-haul, pulley-hauley, are expressions for continually bracing the yards about in variable weather. They are used in shore speech to characterize political maneuvering or vain argument.

(2) To propel a boat by oars—a term consistently preferred by sailors and coastal people to row, and often used intransitively: "Let's pull across to the **Bar** and have a clambake."

In this sense, the word is common in shore speech: Pull for the shore—make for safety; a long pull and a strong pull—a determined effort; let's all pull together—cooperate; to pull one's own weight—do one's fair share. Whether to pull through—recover from a serious illness or surmount some almost hopeless obstacle—refers to rowing, is not known.

Pump ship. To rid the ship of water by means of the pumps. To urinate.

Purchase. In the sense of leverage, purchase belongs entirely to the sea. Any sort of **tackle** is a purchase. Alongshore, the word is used to mean almost any sort of grip or leverage. "When you heft that cask, be sure and get a good purchase on it."

Purser. See **Handspike.**

Push off. Navy slang adopted into shore slang, meaning to leave for, depart; from the order in pushing a boat away from a wharf. See **Shove off.**

Put. A word having various special uses in combination to indicate movement of a vessel. She puts about, or is put about, from one tack to the other, by putting her helm in the direction ordered.

When in distress, she puts back (to the port she left being understood, not generally stated), or puts in to a port of call. (For the phrases used when no distress is involved, see **Call at, Touch at.** See also the discussion on page xv.)

None of the foregoing phrases have found any particular currency in coastal speech, but to put out or put out for, meaning to leave port, has left its mark, in phrases meaning to start off briskly. "The bell rang, and he put out for school lickety-split." An imperative is "You put for home!" or simply, "You put her, now!"

Quarantine (Pronounced currenteen). The period during which ships coming from infected ports are held incommunicado; also, by extension the station of the port health officer who inspects incoming ships. The word refers to the period of forty days covered by the earliest quarantines, which were probably applied to ships. If this was the case, shore uses of the word come from sea experience. See **Yellow Jack.**

Quarter. The after portion of a vessel, forward of the stern and abaft the **beam.** Anything on, or off, the starboard or port quarter bears about 45° from the ship's axial line.

Quarter-deck. The after part of the deck—the domain of the officers. (The forecastle head used to be called "Jack's quarter-deck.") The term is occasionally used to indicate haughtiness—a quarter-deck manner.

Quartering wind. See **Wind** (2).

Quarters. Stations in battle on a warship. To call (or beat) to quarters, that is, to sound an alarm, is sometimes found in literature without reference to the sea. See **Close quarters.**

Quintal (Pronounced kentle). A fisherman's weight of 100 pounds. Permanent ballast of pig metal placed near the keel is kentledge—that is, "quintal-age."

Raft of, a. Shore slang meaning a great quantity of. It may originate in the large rafts of deals which are floated off to ships to be loaded as cargo, in British-Canadian and Baltic ports.

Raise the wind. See **Wind** (1).

Rakish. Saucy; swaggering; from the appearance given a ship by slanting or raking her masts toward the stern. When, however, the adjective is used to mean lecherous, loose-living, it probably comes from the 18th century slang term "rakehell"; but the two meanings may have become confused.

Rammerees, the. The Diego Ramirez rocks, a dangerous hazard off Cape Horn.

Ranzo. Short for Lorenzo: a name applied to men from the Azores, of whom there were many on whale ships. "Oh, Ranzo was no sailor, so he shipped on board a whaler," as the old shanty puts it.

Rapping. Used as an augmentative, this word is probably of nautical origin. It is now becoming obsolete, both at sea and on the land. Sails were said to be "rap-full"; a former shipmate would have a rapping good berth ashore. It may be connected with ripping—a ripping good time, etc.

Rate, to. Naval slang taken ashore; from ranks in the Navy called ratings. In shore speech, it means to have earned, to deserve. "He rates a good licking." See **First-rate.**

Rats. See **Ship** (1).

Rattled down. Having the ratlines, a part of the ship's rigging, repaired and in good order.

Reckoning. Calculation of the ship's position. When unable to get a **sight** on sun, moon or stars, the mariner went by dead reckoning, i.e., by estimating the speed and direction taken. (In this instance, dead probably comes from the abbreviation ded. meaning deduced. See **Dead** for its customary uses.) If the landfall proved him wrong, he had been out of his reckoning, or had missed (lost) his reckoning. All these phrases have crept into shore speech, dead reckoning to mean guess or intuition, the others to mean at a loss, out of one's depth, or simply mistaken.

Reef. (1) A shoal just above or below the surface of the water. The word appears in many figurative phrases ashore—reefs and shoals, the reef one struck on, etc.

(2) To lessen the area of sail exposed by tying part of it to the spar with reef points. Alongshore, to let out a reef is to loosen clothing after a hearty meal, while to take in a reef means to modify one's conduct, "sing small," or be more economical. To be reefed down is to be prepared for difficulties; while always reefed down and standing on the inshore tack indicates the last word in caution!

Reefer. An officers' watch coat used at sea; the name is applied ashore to a short coat worn by children.

Reef knot or **Square knot.** A knot used in reefing sails, and for countless other purposes on shipboard. Its virtue is that it never slips, but can be easily untied. The First Aid Manual of the American Red Cross insists on reef knots for tying bandages. See **Granny knot.**

Reel off. See **Log** (1).

Reeve. Says Captain John Smith in 1625, "reeving is... drawing a rope thorow a blocke or oylet to runne up and down." No better definition can be supplied today. The term is applied ashore to operations involving threading. The correct past is rove.

Regular-built (Pronounced reg'lar). A term of approval, originating in the ship yards, but frequently heard in shore speech.

Reprisal. See **Letters of marque, Privateer.**

Rest on one's oars. See **Oar.**

Ride. To lie at **anchor,** or hove-to. See **Heave to.** After anchoring, the order dismissing the crew is "Let her ride." This reappears in shore slang as "let it ride," i.e., don't pay any attention, "skip it." Figuratively, to ride (out) the storm (gale) is to survive difficulty and danger. For tide-rode, also wind-rode, see **Tide.**

Ride down. To master a sail in furling. Said of a person, it means to domineer. "He rode me down like you'd ride the spanker."

Riding-light. See **Lights.**

Rig. (1) The classification of vessels according to the number, shape and position of spars and sails—ship rig, brig rig, etc. Alongshore, and to some extent in shore speech, the term is applied to dress, clothing. "That was a queer-lookin' rig the new teacher had on to meetin'." A horse and buggy used to be called a rig, probably from this same source.

(2) To fit a vessel with sails and cordage. To rig up is to contrive ingeniously; rigged out (fit to kill) means dressed up, overdressed.

Rigging. Cordage. Standing rigging includes the heavy stays which have only occasionally to be **set up;** running rigging is that constantly in use. The 'longshore proverb, "Old standing rigging makes bad running gear"—advice not to waste labor on worn-out materials—points up the difference. A village philosopher once remarked, "If ye leggo the standin' part of a five-dollar bill, it'll all unreeve before ye can stop it."

Rise. The only use of rise as a transitive verb occurs on ship-board, in such orders as "Rise tacks and sheets" (preliminary to tacking ship). This phrase appears occasionally alongshore as a prompting to hurry up. See **Tack.**

River, the. The River Plate or merely the River, means Rio de la Plata and its ports—Buenos Aires, Rosario and Montevideo.

Road, Roadstead. A sheltered stretch of water where ships may ride at anchor—as at Hampton Roads, and at Royal Roads in Puget Sound. This is the earliest sense in which road appears in the English language. Weekley says that it did not replace way or street until the time of Shakespeare, who first used it in speaking of the land.

Roaring Forties, the. A region of heavy winter gales in the North Atlantic, lying roughly between the 40th and 50th parallels of latitude.

Rock, the. The Rock of Gibraltar.

Rock. (1) Rock in the sense of a sea danger enters into several shore expressions—to go (or run) on the rocks, implying usually financial disaster; the rock one split on, or the cause of a disagreement or mishap.

(2) Rock as applied to the motion of a ship is a landlubber's word. She rolls, or she pitches—sometimes she does both, and then she corkscrews! But she never rocks, in spite of that basso's delight, *Rocked in the Cradle of the Deep.* Even the common injunction "Don't rock the boat," is a landsman's and not a seaman's expression. See **Tip. Rolling down to St. Helena** (Pronounced Aleena). The sailor's term for pleasant sailing weather in the **Trades.** In coastal communities, it means having a fine time in general. "Here comes Dan, rolling down to St. Helena with both sheets aft and stu'ns'ls set" means that Dan, with a few drinks aboard and a girl on either arm, is strolling down the street.

Rope. "There are only seven ropes on a ship," the old saying goes—most of them are lines. See **Line.**

Many phrases current ashore come from a vessel's cordage, however. To learn (or know) the ropes appeared about 1850. Others are at the end of one's rope, get a taste of the rope's end, and the figurative, ropes of sand. (But roped in probably comes from the cowboy's art.)

Rostrum. A rather pompous term in modern journalese for a speaker's platform. This is a very old borrowing from the sea. The rostrum was the Latin name for the bronze beak or ram on early war-vessels of the Mediterranean.

In 338 B.C. the Romans captured an enemy fleet at Actium (which is Anzio, where our troops gained a beach-head in February, 1944). This marked a great advance in Roman sea-power, and the rostra of some of the captured vessels were taken to Rome and set up as trophies in front of the speakers' stage at the Forum. Hence, the whole stage came to be called the rostra. Modern usage has changed it to the singular, rostrum, for some reason unknown.

Rote. A peculiar dull roar made by waves breaking on a beach.

Rough it, to. To suffer hardships—a term created by seamen in the 18th century. Its sense has become softened on shore lips, to mean living a rugged outdoor life.

Round robin. This is properly to be defined as a request or complaint on which the signatures radiate from a center, so that no one can be picked as the ringleader. The device was first used by English naval seamen in the 18th century.

Landspeople use it as the name of a letter which circulates among a group of friends, each adding to it and passing it along to the next.

Round turn. See **Bring up, Hitch** (2), **Turn.**

Roustabout. Originally a deck-hand on river steamers; later used in shore speech to mean a common laborer, circus hand, and so on.

Rover. The original name for a pirate; cognate with the land "reiver" or cattle thief. Ashore, it has come to mean merely a wanderer. In croquet, it means a player who refrains from going "out" in order to assist a lagging partner.

Roving commission. Authority given a naval commander in time of war to cruise wherever he sees fit. See **Commission.** The term is used in many connections by landspeople— equivalent to "free-lancing."

Row (a boat). See **Pull** (2).

Rudderless. A landsman's term, meaning purposeless, without direction.

Rules of the road. See **Collision, Lights.**

Rummage. (From the old French word run; Teut. rum, run, the hold of a vessel.) Its earliest meaning was to stow cargo; it later came to be applied to the clutter and confusion of goods in process of stowage. In shore speech it means to ransack, to disarrange while searching.

Run. (1) The distance covered by a vessel in a given time. (The sailor in a restaurant who complained that the oysters in his stew were a day's run apart was using the word in this sense.) A crew engaged for a single trip are said to have shipped for the run, while a sail allowed to come down of its own weight instead of being carefully lowered, is said to come down by the run.

The word has been completely incorporated into railroad jargon—the run from New York to Chicago, for example. The shore phrase to have a run-in—that is, to quarrel with someone— may come from this source. To keep run of is undoubtedly of sea origin.

(2) As a verb, to run means to sail, especially to sail off the wind, not close-hauled, but running free. By extension, it means to navigate and manage a vessel. In this sense, it is used ashore to cover managing or operating anything—a typewriter, a church social, an armament plant. This is "a definitely nautical use of the word," according to Chase.

This nautical run appears also in many combinations in shore speech—to run across, or find accidentally; to run in with, or meet by chance; to run onto a piece of information; to run counter to a current trend. Perhaps the most directly nautical by inheritance is to run before the wind, that is proceed fast and easily. For other uses, see **Afoul, Aground, Astern, Blockade, Course, Cut, Rock** (1).

Run down. To be in collision with another vessel by running into her. In shore speech, the term means to disparage. To be run down, that is, to be physically under par, may also be connected.

Ryo. Pronunciation of Rio in place names.

Sagged. Of a vessel, sunk in the middle—the opposite of hogged, q.v.

Sail(s). (1) A ship's **canvas.** When under all plain sail, she has all her usual sails set. Seamen have many slang terms for sails—kites, muslin, washing, and for the hated small fancy pieces, "ladies pocket handkerchiefs." They do not, however, speak of a ship as being "dressed" in her sails—that word is reserved for flag decorations. See **Signals.**

A sail is also another ship sighted. This probably goes back to the days of the Vikings, when a vessel carried only a single sail. For a plural form, see sail of **vessels.**

Alongshore, a strange sail is a person seen approaching who is unknown to the speaker. "Cut your sails to fit your canvas" may be either advice to be prudent, or a sardonic comment on over-cautiousness.

The process of handling sails has contributed many phrases to shore speech. To **make** sail for somewhere is to start off in a hurry; with all sail set, full sail ahead, under full sail, or under press of sail means proceeding fast, determinedly; see **Carry on, Crowd on.**

To trim one's sails according to the wind is to be guided by expediency; this is the origin of the slang phrase "a trimmer." To take in, or shorten, sail is to proceed more cautiously. See **reef.** "After you get past sixty, it's time to shorten sail."

To take the wind out of one's sails is to sail windward of another vessel and thereby cut off the wind; hence to frustrate or forestall.

(2) In the seaman's vocabulary, the verb to sail is pretty much confined to the act of leaving port. So strong is the pull of this usage that steamship men have never been able to legitimize the verb "to steam"—they too continue to sail when sailing day comes around. Only a shipmaster, as the man who directed the

operation, says "I sailed"—from such a place for another. The rest of the ship's company modestly says "We sailed."

For the vessel's performance, sail is used in some connections. She sails well or poorly—is a good sailer or a bad. When clear of debt, she sails on her own **bottom.** A day's sail is the same as a day's run. But in general, other verbs are preferred; see **Make, Stand.** One goes sailing only in a pleasure **boat.** The Nantucket whaling captain who arrived home after a two-year voyage, reporting "We hain't got ne'er a barrel of ile, but we had a damn fine sail," was deliberately using summer visitors' language to heighten his effect.

In shore speech, to sail in is to attack someone or something with vigor: "Sail in and win!" A stately lady sails into a drawing room (see also **Sweep**), Smooth sailing, clear sailing, and sailing before the wind are self-explanatory, while to sail near, nigh, or close to the wind may mean either to be economical, or to live up to one's means. See **Plain** sailing, sail under false **colors.**

Sailing orders. Self-explanatory. Ashore it may mean orders to begin anything, or dismissal, jilting.

Sail loft (Pronounced sai'loft). A large open room where sails are manufactured alongshore.

Sailor. This is a much later term than **seaman** (which in turn was preceded by the now obsolete word shipman). See also **Mariner, Jack.** Sailor is also somewhat more restricted in meaning, being applied only to foremast hands, and often preceded by the disparaging word common. His own term of disparagement is more vigorous: "a dirty dog and no sailor." See **Holy sailor.**

A man adept at knots, splices and the like is a marlinespike sailor—one who practices "sailorizing," which is making sea fancywork. "Joshaway made me this lamp mat; he's a master hand at sailorizing." See **Scrimshaw.**

Salleeman. See **Portygee.**

Salt, an old. An old sailor, but probably a shore name from the beginning. When sailors use it, you can hear quotation marks around it. See **Mattalow, Shellback.**

Salted down. A fisherman's phrase, taken into shore speech to mean saved or hoarded. "I imagine he's got a nice little pile salted down."

Salt horse, Salt junk. See **Horse.**

Salvage. In admiralty law, a claim on the value of a vessel and cargo abandoned at sea, by another vessel that has brought them to safety. In shore speech, it is applied to worn-out or discarded material which can be worked over into something of value (very familiar through the recent war salvage drives).

Samshoo. Ardent spirits of any kind; pidgin English brought home by sailors.

Sangaree. Spanish sangria; a West Indian drink of spiced and sweetened wine, introduced by sailors and now used by shore people.

Sargasso Sea. An imaginary region of the sea, known only in the literature of the land (although sailors may have been responsible for starting the yarn) where derelict ships float on through the ages. Actually, the Sargasso Sea is a region of the central Atlantic where, through some peculiarity of currents, floating drift-weed accumulates in large fields, which, however, are entirely navigable.

Savvy. Spanish sabe, via pidgin English, brought home by sailors. He savvy plenty (got plenty savvy) means he is shrewd; no savvy is I don't know; no savvy nothing means that the person spoken of is a fool.

Scanderoon. A kind of carrier pigeon; it received its name from Iscanderun, a Syrian seaport whence news of the arrival of vessels was transmitted to Aleppo by pigeon post.

Scandihoovian. A later name for a squarehead, q.v.

Scarf, Scarph. A carpenter's term for a joint made by beveling and bolting together two pieces of wood in such a way as not to increase the thickness. The name and the process were the invention of ship carpenters.

Schooner. (Probably a Dutch word originally.) The name of a fore-and-aft rig; but landspeople tend to apply it to anything under sail. Alongshore, schooner-rigged means rather poorly dressed. See **Square-rigged;** also **Just.** The heavy glass beer mug called a schooner may come from the nautical word, but it is difficult to imagine how.

Scoop. To shovel, spoon, or ladle. This was originally a nautical word taken over from the Dutch, like **dredge,** q.v. All shore uses of the word come from its use at sea.

Scoot. From the Dutch schuyt, to sail fast. Originally a sea word, this is now completely a shore colloquialism. It survives alongshore as "skite," with an approximation of its original Dutch pronunciation. "I'll have to skite along now; it's time the mail was out."

Scowegian. A late name for a Swede; see **Squarehead.**

Scrimshaw. Fancy articles carved from bone or ivory, usually by whalemen on long voyages, and found on every 'longshore whatnot. British seamen call it scrim-shander-work.

Scud. This originally referred to the running of a hare; it soon became chiefly nautical, meaning to run directly before a heavy wind with scarcely any sail set, or under bare poles. Whether its shore use, to mean running very fast while bent over near the ground, comes from the sea use, or goes back to the hare, is not known.

Scull. To propel a boat by means of a single oar at the stern. Alongshore, to scull around is to be busy in a somewhat aimless fashion. How the term got transferred to the scoop-shaped oars used by racing oarsmen is not clear.

Scuppers. Outlets through the bulwarks for water from the deck. Alongshore, full to the scuppers means intoxicated. "There'll be blood in the scuppers," a term not unknown to landspeople, predicts violence of some sort. The British scuppered, meaning killed, wiped out, probably comes in some obscure way from the idea of a ship sunk till her scuppers are awash, when nothing can save her.

Scurvy. A disease due to vitamin deficiency which used to rage on shipboard. The disease is now rarely seen, but the word persists as a disparaging term in shore speech, a "scurvy trick."

Scuttle. (1) A **hatch** cover, particularly one that closes with a lid sliding in grooves.

(2) To sink a ship intentionally, by those on board. This may be a criminal act, or one perfectly legitimate, as when it is necessary to put out a fire in the cargo when in port. In its shore use, meaning to destroy wantonly and from self-interest something

that should have been preserved, only the unsavory aspect has been retained.

Scuttle butt. A cask of drinking water equipped with a scuttle, which stood on the deck of old-time vessels for the convenience of the crew. It was a good place to exchange views, as men waited their turn; hence scuttle butt, meaning rumor and gossip, a term which has been revived recently in our Navy, and is attaining some currency in shore slang.

"Between you and me and the scuttle butt" is the seafaring equivalent of the landsman's "between you and me and the bedpost." See **Tom Cox, Windlassbitts.**

Scylla and Charybdis, between. A shore phrase, meaning amid insurmountable difficulties, on the horns of a dilemma. It refers to the classical legend of a rock and a whirlpool opposed to each other in the narrow Strait of Messina, each haunted by its appropriate demon, which lurked there to make their prey any unwary mariner who ventured near them.

Sea, the. As with **ocean,** q.v., the sailor avoids the general use of "the sea." In nautical dialect, only ships, never persons, put to sea. A person goes to sea, or follows the sea as a career. "Who wouldn't sell a farm and go to sea?" is the seaman's sarcastic comment during a gale of wind, but, in the language of one old sailor, "he fergits it the first fine day."

Alongshore, a common expression is in all *my* goin' to sea, meaning in all my experience. Many shore phrases make use of the way of a ship upon the sea, such as all at sea, a sea of troubles, and "betwixt the devil and the deep blue sea." For sea meaning **wave,** see **Breakers, Ship** (3).

Sea bag. A sack of sewn canvas, used by sailors and lumbermen as a substitute for a trunk.

Sea biscuit. A very hard cracker. See **Hardtack.**

Sea boat. A ship's quality of performance in heavy weather, and how well she rides out storms, is expressed in the phrases a good sea boat and a bad sea boat.

Seaboots. Hip-length rubber waders, used at sea and alongshore. A face like a seaboot is one much wrinkled and gnarled.

Sea chest. A wooden box, larger at the bottom than at the top, in common use alongshore as well as at sea.

Sea cook. See **Son of a sea cook.**
Sea-glin or **glen.** A light spot on the horizon on a stormy day.

> See a sea-glin,
> Catch a wet skin.

Seagoing. Navy slang, in limited use ashore, meaning competent, or of first-class quality.
Sea lawyer. An officers' term for a sailor, who gives them an argument or serves as spokesman for the crew. It is occasionally used by shore people to mean an argumentative person, a casuist.
Sea legs. Ability to compensate for the vessel's motion in walking. Used jocularly ashore in the case of intoxicated people.
Seaman. The sailor's professional term for himself. See **Able seaman, Ordinary seaman. Sailor** is considered rather a landsman's word—as in the somewhat derisive term "sailorizing."

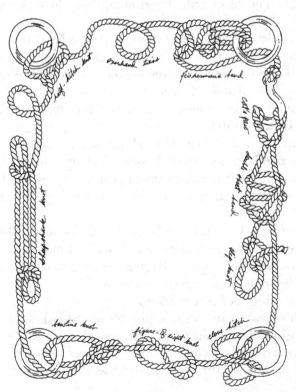

A term of praise, applicable equally to officers and men, is a good seaman. Seamanship covers all skills in the processes of handling a ship at sea. See **Jack Tar, Mariner;** also **Navigation.**

Sea-room. Room to navigate or maneuver a vessel. Alongshore, it is used in a similar sense. "I've got to have more sea-room for this business."

Seat of Ease. A privy; from the convenience on the over-hanging stern of a pinky, a small coasting vessel. (Cf. the British term, chapel of ease.)

Seaworthy. Of a vessel, sound and fit to be sent to sea. Alongshore, it means sound and adequate for any purpose.

Seize (Dutch seizen). To finish off with a wrapping of cord, as a rope's end, a laniard, or the handle of a tool. In this sense, the term is of nautical origin. Alongshore, a necktie or muffler is sometimes facetiously referred to as a "throat-seizing."

Sennet, Sinnet. Flat braided bands of rope.

Set. (1) The fit of sails. Used alongshore in same sense: "I don't like the set of those sleeves."

(2) To hoist a sail and sheet it home. Alongshore, to set sail for is to start off for a given destination. One sets an umbrella or parasol. Everything set and drawing means that all is proceeding favorably. See all **sail** set. (But the shore slang phrase "All set!" refers rather to the position of runners at the start of a race.) See **Adrift, Colors, Furl.**

Set up. To make **taut;** get in all the **slack,** said especially of standing **rigging.** See **Swig.** It is used alongshore in the same sense; also in the phrase, to set (a person) up by the lifts and braces, which means giving a searching and dispassionate criticism of his conduct.

Rays of light from the sun behind a cloud—called by lands-men, the sun drawing water—are described alongshore as the sun setting up his backstays. In shore speech, a person of erect, athletic figure is described as well set up.

Shake a leg. See **Show a leg.**

Shallow water, in. A shore phrase; see **Shoal.**

Shanghai. To kidnap or coerce against one's will; from the abuses practiced by boarding-house keepers who put drugged or drunken men aboard ships to serve as sailors. See **Crimp.**

Shank painter. A short rope used in catting the anchor.

Shanty. A **work**song used to aid labor at sea and sung for pleasure alongshore. (The name is probably derived from the Maine woodsmen's shanty or bunkhouse; derival of chantey from the French chantez is "an educated blunder," according to Philip Barry, authority on folksong.)

Shares. (Often pronounced sheers.) See **Vessel property.**

Shark. A name bestowed by the sailor upon the sea-creature which he hates most—the voracious, cowardly, implacable fish whose triangular dorsal fin, dogging his vessel for days sometimes, gives him the creeps! Whether this is a coined word, or derived from some shore term, is not clear. Men of the press-gangs were sharks to the 18th-century seaman.

Hungry as a shark is a common simile among landspeople. See also **Landshark.**

She. It is not true, as commonly supposed, that the seaman has always regarded his ship as feminine. In the 16th and 17th centuries a ship was masculine—man-of-war, Indiaman and merchantman are cases in point.

Nowadays, she, and her and hers are used whenever reference is directly to the ship in the speaker's mind; but if he happens to be thinking rather of the people aboard, and what they are doing with the ship, he slips easily into he, him and his; sometimes with very mixed results. "Why don't he head her up? She won't weather the point unless he hauls his wind. She needs her **to'gannels'ls**—There, she's hoisting 'em now."

Sheath knife. A blunt-pointed, general-utility knife carried by seamen in a leather sheath, and in use alongshore as well.

Sheave (Pronounced shiv). The moving part of a block or pulley.

Sheepshank. A temporary knot used to take up the slack in a line.

Sheer. The curve or rise of a ship's decks and bulwarks, from bow to stern. Used in the same sense of objects alongshore. "Seems like the back of that sofa hasn't got enough sheer to it."

Sheer off. To change course away from an object. Ashore, the expression means to draw aside, or to change the subject. See give a wide **berth.**

Sheet. As used on shipboard, sheet at first meant sail; it has now come to mean one of the principal ropes by which a sail is handled, and it constitutes a really heinous error on the part of a landsman to refer to the sails as sheets.

To sheet it **home** to a person means to prove him delinquent, or to force him to shoulder some responsibility that is rightly his. (Partridge, who is no sailor, says "perhaps from an entry of a person's name on a charge-sheet"! These etymology boys are not to be relied on implicitly.)

Three sheets in the wind, a sailor's phrase meaning intoxicated, is frequently used by shore people. Both sheets aft also describes a sailor feeling the better for a few drinks, and on top of the world. See **Pay off** with the topsail sheet.

Sheet anchor. See **Anchor** (1).

Shellback. The sailor's commendatory term for the landsman's old **salt,** q.v. See also **Mattalow.** Some authorities say that it comes from his back being bent like a shell; but it seems more probable that the implication is that the shellback is growing barnacles from having been at sea so long.

The term is fairly well known to landspeople. The degree of "able shellback," signed by Rex Neptune, is currently being conferred upon men crossing the Line for the first time aboard troopships.

Shift. Sailors prefer this word to change, move, or alter, whether referring to bending a fresh suit of sails or changing their own clothes; and 'longshore people follow the same preference.

To shift over—change position—as used by shore speakers, is undoubtedly of nautical origin.

Ship. (1) Originally a three-masted vessel **square-rigged** on all three masts, but also used as a general term for all square-rigged vessels, and now extended to include steamers when used in some connections.

Alongshore, in charge of the ship means managing or directing an enterprise; to keep ship is to serve as watchman, or to look after the house while the rest of the family are out. "What ship is that?" is a sly hint that the speaker is using some long word that the inquirer does not understand. To be out of a ship is to be unemployed; and this phrase has been reported to be current among people in the theatrical profession to mean the same thing.

Shore people make use of the ship in many figurative phrases, among them airship; the Ship of State; ship of the desert (a camel); as the rats desert a sinking ship; enough to sink a ship; don't give up the ship; and when my ship comes in. See **Boat, Vessel.**

(2) To forward as cargo in a ship. This word, together with its cognates, shipper and shipment, has now been taken over completely into the language of trade and transportation ashore.

(3) To take on board of a boat or vessel—either voluntarily, as cargo, or involuntarily, as a sea. Alongshore, to ship a sea means also to get one's feet wet. A landsman's perversion is to ship one's lunch—i.e., vomit.

(4) To hire or be hired to serve in a vessel. Figuratively, to "ship for the voyage" means to get married.

(5) To fit into a prepared setting, as oars, or a rudder. See **Step.**

Shipmate. A person who sails or has sailed with one in the same vessel. It is also extended to objects: "I've never been shipmates before with a new-fangled machine like that."

Shipping. Ships and boats collectively; anything that moves on the surface of the water.

Ship's company, the. Same as all hands. See **Hand** (1).

Shipshape. A sailor's commendation of anything in first-class order or condition—the superlative being shipshape and Bristol fashion. Shipshape is quite common in shore speech, and bears the same meaning.

Ship's husband. A business agent for ships in port; a term understood and used alongshore.

Shipwreck. Complete disaster to a vessel. The word is used figuratively ashore: "The shipwreck of his plans caused his health to break."

Shoal. A place with little depth of water; almost a reef. In shore language, to be in shoal water means in a dangerous or ticklish situation; reefs and shoals are figurative perils.

Shoals, over the. This phrase always refers to Nantucket Shoals.

Shoot the sun. To take an observation of the sun to ascertain the vessel's position. The phrase is used in airplane navigation. See Sight (1).

Shore. This word is used by sailors chiefly as an adjective—shore people, shore (or shoregoing) clothes—or in combinations: ashore, offshore, on-shore, inshore.

Shorten. See **Sail** (1).

Shot. Of the anchor chain, a length of 15 fathoms. Another shot in the locker means that there is more chain ready to stick out if occasion demands. The phrase has come ashore to mean something in reserve, still to be drawn upon. Its users have no idea of its meaning. See **Bitter end.** Not by a long shot, probably also a term of sea origin, relates, however, to a gunshot. See not by a long **sight.**

Shove off. To push a boat away with the oars, hence, to start away. The term has come ashore via the Navy, and been incorporated into American slang. See Push off.

Show a leg. A hail formerly used in calling the watch below. It is said to come from the time when women were allowed aboard ships of the British Navy (see **Son of a gun**), the object being to permit the boatswain to see, from the nature of the hosiery displayed, whether a man or a woman occupied the hammock. As the term ceased to have any significance, "Shake a leg!" was substituted; and this has passed over into shore speech to mean make haste.

Sick Bay. The hospital on a naval vessel. In sick bay is sometimes heard ashore, in a figurative and facetious sense.

Side lights. See **Lights.**

Sight. (1) An observation of sun, moon or stars, taken to calculate a ship's position (called working up the sight). The shore phrase not by a long sight probably originates in this practice. See not by a long **chalk, shot.** In sight has no special relation to the sea; but to **heave** in sight is distinctly nautical in origin.

(2) To discern; see **Make** out, **Pick up.** Any use of to sight for to see is nautical in origin—e.g., to sight trouble ahead.

Sign. To sign the ship's crew list at the beginning and ending of a voyage, called signing on and signing off. See **Articles,**

Ship (4). Alongshore, these phrases mean taking or quitting a job. "I didn't sign on to run errands."

Signals. This word usually refers at sea to code flags by means of which long conversations can be held between ships proceeding on the same course—a practice known as "signalizing." To celebrate a holiday in port, all the code flags are set; the ship is then said to be "dressed." For signals of distress, see **Distress, S O S, Union down.**

Alongshore, a person waving to attract the attention of someone else, is said to be signalizing.

Sing out. (1) To hail or call loudly.

(2) To chant while pulling on a rope. In shore speech it means, to speak up. "If you need anything, sing out." See **Yeo-heave-ho.**

Sink. Of a ship, to go to the bottom. See **Swim.** The shore phrase sink or swim, i.e., be dependent entirely upon one's own resources, comes from seafaring and not from natation. To sink money in something is probably nautical in origin; hence also sinking fund. Sunk, meaning exhausted, overcome, borne down by adverse circumstances, also harks back to the ship. The oath "sink me!" now obsolete, was popular with London bucks in the late 1700's. See **enough to sink a ship.**

Sir. See **Forms of address.** Alongshore, the old sir means a retired shipmaster. "The old sir picked him a sightly spot for a house, didn't he?"

Siren. A mythological creature, half woman and half bird, who was believed to haunt certain rocky isles in the Mediterranean and, by her sweet singing, lure mariners to destruction on the rocks. See **Mermaid.**

In modern speech, a siren is an alluring and somewhat dangerous female.

Skipper. (Dutch schipper, pronounced skipper). A familiar term for the master of a small craft (to be used with caution; see **Forms of address**). Ashore, it is used in many connections for boss, head man. Airplane pilots are called skipper. A fashionable pet is the schipperke, "little skipper," a small Dutch barge dog.

Skoff'm. A sailor's term for food in general; from the Swedish skaffning, meaning grub. Picked up at sea from Scandinavian sailors, and brought home to the coast, where it even appears as a verb: "Skoff it up quick, now!"

Skylark. A sailor's term for romping, chasing one another up the rigging, etc. Lark is the old English, lake, to sport or play. Skylark is now common in shore speech.

Sky pilot. The sailor's name for a preacher, particularly a missionary to seamen or a naval chaplain (although I am told that the latest naval slang for this officer is "fire escape").

Skyscraper. A small triangular sail carried on some ships above or in place of the skysail. Ashore, it means a very tall building. This (like **cutwater,** q.v.) may be merely coincidental—a case of multiple naming.

Slack. Rope with no strain upon it, so that it is easily pulled in. The verb form is usually to slack, not slacken.

Alongshore, "Slack away!" is the ordinary command to let something out slowly. Holding on to the slack is just about holding one's own. "He ain't good enough even to hold the slack" is self-explanatory. Figuratively, slack is back-talk, impudence; while slack-jawed means garrulous or foulmouthed: "He's nothing but a slack-jawed chaw-mouth!" A very hungry man will say he can take up the slack of his belly and wipe his eyes with it.

The only shore phrase in common use from this source is to take up the slack of something. (To give a person slack line probably comes from sport fishing.) See slack **water,** slack in **stays, Taut.**

Slant. A leading wind; see **Wind** (2). Ashore, it means a bit of good luck.

Slatch. A let-up in a spell of bad weather. Figuratively, a respite, a bit of free time. "I've got a little slatch this morning and can help you out with that job."

Sleep in. To sleep through one's watch on deck (if the mate didn't get on to it!). To sleep all night. In one section of Ohio, this term is habitually used for sleeping late in the morning.

Slew. To change course; to **yaw.** Originally a nautical word, from the Dutch. All shore uses of the verb are believed to come from the sea use.

Slick. An oily-looking streak on the water, caused by a calm patch amid flaws of wind—the opposite of a catspaw.

Slumgullion. Originally, entrails and refuse after a whale had been "cut in." A vulgar term, at sea and alongshore, for a meat stew.

Slush. Waste fat from the galley, used to grease the masts. It used to be the cook's perquisite. Ashore, a slush fund means money that may be used for bribes with no account given; graft.

Smack (Dutch). A small fishing vessel equipped with a salt-water tank for keeping fish or lobsters fresh (Wasson). It is used erroneously by shore people for any fishing-craft, even the big Glo'stermen.

Smelt, smooth as a. Sleek and handsome.

Smoke and oakum! A meaningless exclamation of astonishment.

Smuggle. To bring goods ashore from a ship without paying the required duties. Ashore, it means (figuratively) to convey by stealth.

Smur up. To become cloudy or overcast.

Snub. To throw a turn of rope around a pin or bitt in order quickly to check it, and used in the same sense alongshore—for example, of a cow's halter.

Snugged down. Same as reefed down, see **Reef** (2). Alongshore, it means prepared for eventualities. "We've got all snugged down for winter."

So! Enough; stop pulling or lifting. Same as **belay** or **avast.** Common alongshore.

So-fashion. Thus. A pidgin English word brought home from China by sailors, and in general use alongshore.

Soft tack. Ordinary loaf-bread. A jocular antonym for **hardtack.**

Soldier (Pronounced sojer). The marines on old warships had no duties connected with working the ship, and were accused by the man-of-warsmen of being idlers. Ashore, the term has come to mean to malinger; to loaf on the job. See **Marine** (3).

Soldier's wind, a. A steady beam-wind, with which a vessel can sail equally well on opposite courses. (Another reflection on the seamanship of the marines—they might be able to get the ship there and back if they didn't have to tack her!)

So-long. A seaman's farewell, being, according to Weekley, his imperfectly understood version of the East Indian word salaam. It is now very common in shore speech.

Sonnywhacks. A good-natured form of address to the ship's boys, and used to boys alongshore.

Son of a gun. This term is said by several authorities to be derived from an old custom in the British Navy of permitting wives and lady-friends of sailors to travel on the ships, with the result that children were sometimes born on board on long voyages. It is in general use ashore as an epithet of scandalized admiration.

Son of a sea-cook. An offensive but meaningless epithet, current at sea and alongshore. The **cook** was the butt of many sailor's jokes.

Soojie-moojie. Washing soda or any cleaning compound. A word probably of Japanese origin, brought home by sailors.

S O S. The code letters used in radio to signal distress at sea. (The modern radio operator's maritime distress call is "Mayday" from the French "m'aider.") In shore speech, S O S has deteriorated to mean any call for help, however trivial.

Soul-and-body lashings. See **Oilskins.**

Sound, the. Long Island Sound or Puget Sound, according to the context; also the narrowest water between Denmark and Sweden is called "The Sound."

Sound. (1) To ascertain the depth of water under a vessel by heaving the **lead,** q.v. See also **Fathom, Plumb.** To sound the pumps is to learn how much water has leaked into the ship by dropping a measured gauge into the pump well.

Weekley states, but Chase disagrees with him, that the land use of sound, as in the case of a surgeon probing a wound, was probably the earlier. The shore phrase, to sound out, meaning to reconnoiter, fish for information, may come from either source.

(2) A whale is said to sound when it dives under after being struck. A whaling-captain's wife found her small son trying to harpoon the family cat with a fork tied to a ball of string. She grabbed for the ball, but the boy sung out "Pay out, Ma, pay out! There she sounds through the window!"

(3) Sounds are the air-bladders of cod and other large fish—a delicacy much relished at sea and alongshore, but almost impossible to procure outside fishing communities.

Soundings. Water shallow enough so that bottom can be found with the lead. A ship is said to be on or off soundings. Rarely, but only in shore speech, out of soundings is used to express bewilderment, lack of direction. In soundings is printers' jargon for having come almost to the bottom of a pile of paper.

Soupa de bool-yon, three buckets of water and one on-yon. A phrase which combines derision of foreign languages and criticism of a very thin soup. Another description of the latter is "so thin you can see bottom at forty fathoms."

Southing (Pronounced to rhyme with mouthing). Progress to the south; the vessel makes southing.

Southward (Pronounced suth'ard, to rhyme with mothered). Toward the south.

Sou'wester. An oilskin hat shaped something like a fireman's helmet, in common use at sea and alongshore.

Spanker-boom tea. A mild laxative used on whale ships, made from an herb gathered in South Sea Islands.

Spanking. See **Strapping.**

Spar(s). Masts, yards, booms, etc. Alongshore, long-sparred and short-sparred were terms to describe people and animals with long or short limbs.

Spar varnish. A hard, water-resistant varnish first used on ships, and now a recognized trade name under which this type of varnish is sold on shore.

Speak. To communicate between ship and shore, or from ship to ship at sea, either orally or by code signals. Alongshore, to speak means to accost, hail. "He spoke me and wanted to know—" See also **Signals, Sing out.**

Spike. Any piece of pointed metal larger than a tack is called a spike on board ship. They go by various names—deckspike, handspike, marlinespike. The shore phrase, to spike one's guns is believed to come from sea, not land, warfare.

Spin a yarn. See **Yarn.**

Splice. To unite two ends of rope by interweaving the strands. Alongshore, to get spliced is to get married. To splice the main brace, meaning to take a drink of liquor, is generally understood and sometimes used facetiously in shore speech.

Spout. Colloquially used in shore language, meaning to talk volubly, to orate. Probably from whaling.

Spread-eagle. To **trice up** a person by his wrists and ankles, preparatory to flogging him. The term is sometimes used by landspeople to describe a person's sprawling flat on his face or back.

Spread of canvas. Sails. The colloquial shore phrase to make a spread, or to parade, make great display, may come from this source or, less probably, from spreading a table.

Spree. A sailor's liberty trip ashore, and especially a drinking-bout. The word is very common in shore speech. Its origin has not been determined; and until research may develop facts to the contrary, it belongs in this list.

Sprung. Of a mast, twisted and cracked. The word is not commonly used in this sense ashore—but a Negro girl at a rummage sale was heard to reject a bright-colored knitted dress because it was "butt-sprung"! Alongshore, a person in the early stages of intoxication is said to be started and sprung.

Spurlos versenkt. Sunk without a trace—the treatment recommended for vessels carrying contraband, in an official German dispatch of the first World War. The phrase passed into the colloquial speech of the day, to mean lost, completely disappeared.

Squall. A vicious gust of wind. "Look out (watch out) for squalls!" is, in shore speech, a common warning of trouble ahead.

Square, Square away. Of the yards, to bring them at right angles to the keel and let the ship run before the wind. Alongshore, to square the yards with another person is to repay a debt. "Here's half-a-dozen eggs and a dite more o' my fresh butter; I guess that squares our yards, don't it?"

In shore phrases of this sort, some of the nautical flavor is lost: one squares up or gets square with the world, in paying debts. An honest person is therefore a square one. To get square with someone, is, however, to take revenge for an injury or to even a score. (Whether to square off—i.e., to put up one's fists ready to fight, is from this same source, has been questioned. On the square is definitely from Freemasonry, not seafaring. Cf. on the level.)

Squarehead, a. A native of Denmark, Norway or Sweden. (Finns are Finns; no disrespectful term is ever applied to them, on account of their reputed powers of wizardry.) See **Scandihoovian, Scowegian.**

Square knot. See **Reef knot.**

Square-rigged. Having square sails suspended from yards which run crosswise rather than lengthwise of the ship. Applied to a person (usually a man) square-rigged means in formal attire or full-dress uniform. See **Fore-and-aft, Schooner**-rigged.

Squeegee. A deck scraper. Ashore, used by window cleaners to wipe water from pane; the instrument used by photographers to remove excess water from prints is also called a squeegee.

Squid. From squit, an old form of squirt. A name given by sailors to the cuttlefish, from its power to eject a black secretion.

Stand. A general word meaning to **head.** See **Run.** A vessel stands to a given compass point; stands in to, or out, off, from a port or the land; stands up to, or away from, the wind; stands up, down, or along the coast. She stands in with another vessel when they are sailing in **company.** The latter phrase is doubtless the origin of the shore phrase to stand in with someone—i.e., be in favor. Stand-offish meaning haughty, supercilious, may or may not be from this source. "All hands stand from under!"—a phrase of warning when something is about to fall—is quite certainly of sea origin, however.

A sailor stands **watch**—only a landsman keeps watch.

Stand by. To remain close to another vessel in distress, ready to render assistance if needed. A standby, something or somebody always to be relied upon, doubtless enters shore speech from this source. "Stand by!" is the regular order at sea to await another order. It is now familiar to radio listeners as well.

Starboard (From steer-board). On or toward the right-hand side of a vessel. "Starboard your helium!" is a jocose order sometimes given alongshore. See **Board.**

Stasia. The island of St. Eustatius in the West Indies.

Stateroom. A room for the accommodation of a passenger or officer, on board ship. The word has been said to originate in the Mississippi River steamers, on which the cabins were named for the states of the Union; but the term antedates this, and goes back to the days when only great dignitaries were accorded a separate state cabin on board ship. On shore, the term has been taken over by the railroads to designate a drawing room compartment on a Pullman car.

Staunch. From the Old French estanche, watertight. Of a vessel, seaworthy, not leaking. In shore speech, the word means sound, dependable—a staunch friend.

Stave. To smash; a sea term in origin. The correct past is stove. To stave along means the same as to **carry on,** q.v. Alongshore, to stave along means to proceed, without regard for the comfort or safety of others. But staver, applied to a vigorous, growing boy, is a term of admiration, mixed, perhaps, with some apprehension. Cf. buster. Of such a one, they say alongshore that he "don't fear the face of clay."

Stay. In the sense of to support, this is a word of nautical origin, the masts being held firmly in position—stayed—by heavy cables called stays. "All shore uses derive from this nautical one; as to 'stay one's stomach,' 'stay one's hand.'" (Weekley). See **Mainstay.**

Stays, in. In the process of going about from one tack to the other. A vessel which does not come about readily is slow (or slack) in stays. Alongshore, this phrase is applied to a slow-moving, awkward person.

Steady! "Steady!", "Steady your helm!" or "Steady as you go!" are orders to the helmsman when the ship is headed precisely as desired.

> South-west and a half west, and steady as you go,—
> We've a long way to travel to Rosario!

a British poet sings.

Alongshore, steady is the word addressed to a skittish horse, or even to a person who appears to be losing control of his temper.

Steer. To direct the course of a vessel by means of a wheel or tiller controlling the rudder. In shore language, to steer is to guide (a conversation, a business deal, an inexperienced person). A steering committee is a small executive body. Modern slang transforms the word into a noun, "a bum steer." The word appears in several combinations: to steer clear of, (as Washington, in his Inaugural Address advised that this nation "steer clear of permanent alliances"); and to steer off shore, i.e., out of danger. See **Course, Helm, Wheel.**

Steerageway. Speed (barely) sufficient to permit holding to a course.

Steeve. (1) to cant sharply upward, as a ship's bowsprit.

(2) To stow, pack tightly, as a ship's cargo. See **Stevedore.** It is used alongshore of any tight packing: "Steeve it in good and solid, now."

Stem to sternpost, from. The stem is the bow; the sternpost is the perpendicular timber at the extreme stern of the ship. Hence, at sea and alongshore, the phrase means completely, thoroughly, all over. See **Alow** and aloft, **Clew to earing, Truck to keelson.** See also to stem the **tide.**

Step. Of masts, to raise and affix firmly at the bottom. Alongshore, it means to set anything into a prepared foundation. See **Anchor** (2).

Stern (Often pronounced starn). The after end of a ship. Alongshore, it means a person's backside. By the stern, opposite of by the **head,** q.v.

Stern chase. "A stern chase is a long chase," a saying used when the situation has got a head start on one, dates from the days of the privateers.

Sternforemost. A phrase current at sea and alongshore, meaning clumsily, or the wrong way to. "He always goes at things starnforemost."

Stern sheets. The semicircular seat at the stern of a rowboat. Alongshore, it is applied jokingly to the back seat of an automobile, as well as to a person's rear: "I gave him a good kick in the stern sheets."

Sternway. Movement of a ship backward instead of ahead. It is occasionally used figuratively alongshore, but is not so common as **headway** and **leeway.** See also **Board.**

Stevedore. From Spanish estivador, a packer. A man employed in loading and unloading ships. The term was probably brought home by sailors, and is now in common use ashore. See **Steeve.**

Stick. Sea and 'longshore slang for mast. "Up stick and cut it," refers to stepping the mast of a small boat, and getting out of there pronto! The British slang phrase to cut one's stick doubtless comes from this source, although it was not the mast but the anchor or mooring-rope that was cut under those circumstances. See **Cut and run.**

Stick-in-the-mud, a. A shore phrase meaning a slowpoke— usually with "a regular old" preceding. Reference is to a boat.

Stiff, Cape. See the **Horn.**

Stocks. The frame on which a ship's hull rests while she is being built. See **Ways, Launch.** On the stocks, meaning in preparation, and off the stocks, signifying a going concern, appear occasionally in shore language in relation to business enterprises.

Stopper. A short piece of very strong rope having various uses, as to relieve strain on the anchor cable when riding out a heavy gale. (Not to be confused with a stop which is a bit of small stuff used to tie up a flag, or a sail when bending or reefing it.)

"Put a stopper on your jaw!" is a forceful way of saying "Shut up!" It is not often heard now.

Storm. This is not a word in common use at sea or alongshore, where gale or breeze is preferred except in the case of **line** storm. It appears in many shore phrases, some with clearly nautical implications—e.g. to face the storm, to fly storm signals, to sail on (or through) stormy seas. See also **Eye of the storm,** to ride out the storm, any **port** in a storm, Weather.

Stormbound. See **Weather-bound.**

Stormy Petrel. A small sea bird; the **Mother Carey's Chicken,** q.v. Ashore, the term is used to describe a person constantly in difficulties, or a trouble-maker.

Stove. See **Stave.**

Stow (Dutch stoewen, to cram, press). To pack cargo into a ship; by extension, to pack, to put away. The word is used in latter sense alongshore and to some extent in shore speech. "Where in tunket have you stowed away my fishing tackle?" To stow away grub is to eat. See stow the gaff.

Stowaway. A person who hides on a ship to secure free passage. The word is commonly understood ashore, and is used to describe the same circumstance on an airship. Also, figuratively, it means an article which turns up where it doesn't belong. See Work-away.

Straight sailing. See **Plain** sailing.

Straits, the. The Strait of Gibraltar. A voyage to the Mediterranean used to be called "going up the Straits."

Strake. A longitudinal timber of a ship or plank of a boat. The word is etymologically related to streak, and the two words are sometimes confused by landspeople—notably by the translators of the Bible, with their "ring-straked."

Strand (Anglo-Saxon, meaning shore). To strand, now obsolete at sea, formerly meant to be aground. In shore speech, stranded is used of a person penniless and friendless in a strange city; or of the community itself when left without industrial resources to support its population.

Strapping. Brawny, well-developed. When used to describe the human physique, the etymologists say that this is "naut."—developed by sailors. They do not say this however, about spanking, which is more commonly heard at sea, and which is applied to ships and breezes as well as to people.

Stream, the. The Gulf Stream. (But the stream; without initial capitals, means an anchorage away from a wharf.)

Strike. A more or less obsolete nautical word meaning to make smooth or level. To strike one's sails, colors, was to take them in. To strike aghast, etc., in shore speech, may be related. See strike **adrift.**

Studding sails (Pronounced stun's'ls). Supplementary sails set beyond the yardarms for added speed. See **Boom.** Both the name and the sail are comparatively recent in origin; the regular Middle English word for a piece added on to another sail is bonnet, which still survives.

Figuratively, the term is associated in the sailor's mind with women. A much dressed-up female is said to carry stun's'ls alow and aloft. But a man with stun's'ls set has a woman on his arm.

Suffer like a thole-pin, to. This refers to the anguished squeaks of the old-fashioned wooden oar-fastenings in a rowboat called tholes.

Sundog. A bright reflection of the sun in a stormy sky, near the sun and in the form of irregular spots.

Supercargo. A business agent who formerly made the voyage to represent the owners. He was neither a passenger nor a member of the crew; and the advent of the telegraph put him out of business. Ashore, the word is sometimes used to mean an unnecessary official; a supernumerary.

Surf. Breakers after they run up on a shore or shoal. A surf-day is one on which it is impossible to land or load cargo in an open roadstead.

Surge. To let out a little slack on a rope that has been pulled too taut; to let the capstan run backward a pawl or two. See **Come up.** "Surge old spunyarn!" used to be a common exclamation alongshore when anything parted under a strain.

As a noun meaning surf or waves, surge is unknown at sea or alongshore, and is confined to literature.

Survey. A consultation which must be called by a shipmaster, in case of serious damage to ship or cargo when in or near port, in order to make an official report and advise him how to proceed. To hold a survey is used in other connections alongshore.

Swab. A mop for cleaning the deck or paintwork. Applied to a person, it was a derogatory epithet, now obsolete.

The word has a specialized meaning in medicine, of a pledget for cleaning cavities or wounds. The common expression, to swab out (or up), also comes from this nautical source.

Swamp. Of a boat, to fill with water and sink. See **Founder.** The shore expression, to be swamped—with difficulties, work, and so on—comes from this use of the word.

Sweep. A long oar, used to propel heavy barges, pinnaces, etc. To sweep (into a room) probably comes from the stately progress of a dignitary's barge, propelled by sweeps. See also **Clean sweep, swept hold.**

Swell. Waves underrunning the surface of the sea without breaking. A term occasionally appearing in literature is ground swell—a forerunner or omen of something to come.

Swig. To sway upon a rope that is nearly "hand-taut." While one man holds the turn around the pin, the rest "sweat it up" to get a few more inches of slack. A purely nautical word, but heard alongshore as well.

Swim. Of a ship, to remain afloat. The correct nautical past is swum, not swam. See Sink.

Swivel. A nautical device which permits either end of a length of chain to revolve separately, or a gun to swing on its pivot. Both the word and the article itself have now come ashore, and find many uses.

A person who squints is said to have a swivel-eye. A swivel chair is a desk chair with a revolving seat.

Tack. (1) The lower weather corner of a sail, and the rope attached to it by which it is handled, are both called tacks. See **Rise** tacks and sheets. A vessel is said to have her "tacks aboard," on the side from which the wind is coming, as in the old song:

> With her starboard tacks aboard, my boys, she hung
> up to the breeze;
> It's time our good ship hauled her wind abreast of
> the old Saltees.

The 'longshore phrase, to talk tacks aboard, means to be garrulous and long-winded. A person says he "won't start tack nor sheet" until something else has first been done.

(2) By extension from this usage, the irregular zigzags by which a ship advances when beating against a head wind are also called tacks. She is on the starboard (or port) tack according to whether the wind comes over her starboard or port bow.

In coastal phraseology, at every tack and turn means inconveniently often—usually said in scolding or reproach.

In shore speech, to take the wrong tack means to be undiplomatic; to be on the wrong tack is to err or misapprehend; to try another tack signifies a new approach or expedient; while to take opposite tacks is said of two people who differ with one another.

(3) To tack is to bring a ship up into the wind and around so as to catch it from the other side, meanwhile trimming the sails accordingly. No better account of this operation was ever written than the sailorly verses of Walter Mitchell, *Tacking Ship Off Shore.*

The weather leach of the topsail shivers,
    The bowlines strain and the lee shrouds slacken.
The braces are taut and the lithe boom quivers,
    And the waves with the coming squall-cloud blacken.

Open one point on the weather bow
  Is the lighthouse tall on Fire Island Head;
There's a shadow of doubt on the captain's brow,
  And the pilot watches the heaving lead.

I stand at the wheel, and with eager eye
  To sea and to sky and to shore I gaze,
Till the muttered order of "Full and by!"
  Is suddenly changed to "Full for stays!"

The ship bends lower before the breeze
  As her broadside fair to the blast she lays,
And she swifter springs to the rising seas
  As the pilot calls, "Stand by for stays!"

It is silence all, as each in his place,
  With the gathered coils in his hardened hands,
By tack and bowline, by sheet and brace,
  Waiting the watchword impatient stands.

And the light on Fire Island Head draws near
  As, trumpet-winged, the pilot's shout
From his post on the bowsprit's heel I hear
  With the welcome call of "Ready, about!"

No time to spare! It is touch and go;
  And the captain growls "Down helm! Hard down!"
As my weight on the whirling spokes I throw,
  While Heaven grows black with the storm-cloud's frown.

High o'er the knight-heads flies the spray
  As we meet the shock of the plunging sea,
And my shoulder stiff to the wheel I lay
  As I answer "Aye, aye, sir! H-a-r-d alee!"

With the swerving leap of a startled steed
  The ship flies fast in the eye of the wind

The dangerous shoals on the lee recede,
  And the headlands white we have left behind.

The topsails flutter, the jibs collapse,
  And belly and tug at the groaning cleats;
The spanker slats and the mainsail flaps,
  And thunders the order, "Tacks and sheets!"

'Mid the rattle of blocks and the tramp of the crew
  Hisses the rain of the rushing squall;
The sails are aback from clew to clew,
  And now is the moment for "Mainsail haul!"

And the heavy yards, like a baby's toy,
  By fifty strong arms are swiftly swung.
She holds her way, and I look with joy
  For the first white spray o'er the bulwarks flung.

"Let go and haul!" 'Tis the last command,
  And the head-sails fill to the blast once more.
Astern and to leeward lies the land,
  With its breakers white on the shingly shore.

What matters the reef or the rain or the squall?
  I steady the helm for the open sea.
The first mate clamors, "Belay there, all!"
  And the captain's breath once more comes free.

And so off shore let the good ship fly;
  Little care I how the gusts may blow,
In my fo'c's'le bunk, in a jacket dry.
  Eight bells has struck, and my watch is below.

Alongshore, "Tack ship!" means turn around. "I made the old coot tack ship, anyway," boasted the hunter who had taken a pot shot at a bear, but missed. **Tackle.** (1) Pronounced taycle. A mechanical purchase of blocks and ropes. According to Weekley, the word is "chiefly nautical from the 13th century," and has been borrowed by the sailors of many other nations, being now completely naturalized in these foreign tongues. It is used in the same sense alongshore.

(2) To tackle (pronounced as spelled) means to undertake with vigor, to attack. Chase maintains that these two are the same word. "Such a difference in pronunciation is a familiar linguistic phenomenon.... 'Tackle' seems to be a shore pronunciation for the noun in the sense of equipment." This second tackle is firmly embedded in shore speech, both as a noun and a verb.

Take in. See **In** (1), also Sail (1).

Tally. To keep track of items of cargo going in and out by means of tally marks, four upright strokes with a fifth diagonal stroke binding them together. As used by scorekeepers ashore, the practice is probably taken directly from this sea use; but students of early human history tell us that this system of counting antedates water-borne commerce, and that it is a symbolic representation of the fingers of the hand, upon which the first counting was done. It is probable, therefore, that the phrases to keep tally of and to tally with—i.e., coincide or agree with— exist in shore speech in their own right, and do not derive from sea experience.

Tally on. To line up in order to grasp; to help pull on a rope, and used in the same sense alongshore. Its shore use is limited to the sport called tug-of-war.

Tap the admiral. It used to be said of a sailor who would drink any liquor no matter how bad, that he would even tap the Admiral. This refers to the old tale of the sailors who tapped the cask of alcohol in which the body of a dead admiral was being conveyed home.

Tar. A preservative for wood, rope and canvas, which has become identified with seamen—a tar, "tarry sailor," and the like. See **Jack Tar.** It enters into many phrases current at sea, alongshore, and to some extent into shore speech.

Don't lose the ship for a pennyworth of tar is a common bit of advice not to be too saving. Weekley says this was originally "spoil the sheep," not "lose the ship," (from the farmer's practice of doing something or other with tar for the welfare of these animals), but this hardly seems to make sense, and the aphorism was at one time, and perhaps still is, in wide use among seafarers.

See **Naval** stores. See also. **Every hair, Knock down.**

Tarbrush, a lick (or touch) of the. According to Admiral Smyth, this was first used to characterize recently promoted officers, whose manners still smacked of the forecastle. Its present use in shore speech is to indicate a suspicion of colored blood. But Weekley, still obsessed by sheep (see **Tar**), says it refers to farm practice.

Tarpaulin (Pronounced tarpolin). **Canvas** used to cover hatches, etc., to keep out water. It used to be tarred, but nowadays is water-proofed with other preparations. (The Middle English word from which it was taken is palyoun, or canopy, related to our modern pavilion.) This fabric is in common use ashore, to cover goods in trucks, or exposed to the weather. In shore use, the pronunciation of the word generally follows its spelling.

Tarpaulin muster, a. A pooling of small change by the crew, preparatory to a joint spree on shore, the coins being tossed into a tarpaulin. It is occasionally heard ashore, in the case of taking up a collection for something.

Tarred with the same brush. A shore phrase, meaning affected with the same evil qualities. Sheep, or ship? Take your choice.

Tattooing. A practice borrowed by sailors from the South Sea Islanders, and still more common among seafarers than among others.

Taut. Not originally a sea term, but now chiefly nautical. It means tight; opposite of the seaman's slack. Used in the same sense alongshore; and figuratively by landsmen—a taut smile, manner, etc. The verb form, at sea and alongshore, is to tauten. See **Set up, Swig.**

Tempest. At sea and alongshore, this means strictly a thunderstorm. Its use to mean gale is literary and of the land—tempest-tossed; the tempest roared, etc.

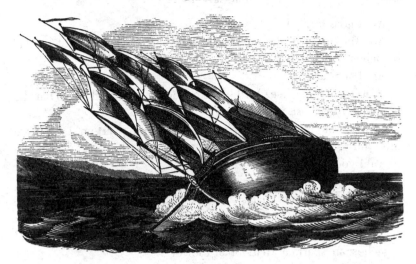

Tender. (1) A small boat serving a larger vessel. Ashore, it is the fuel car attached to a locomotive.

(2) Tender, as an adjective, means the same as **crank,** q.v. and is used in the same sense alongshore.

There she blows! The conventional hail used aboard a whaler when a whale is sighted. It is sometimes used jokingly alongshore.

At the risk of being overlengthy, we must take time out to insert here New Bedford's Whaling Classic, as recorded by the late Zephaniah Pease in his *History of New Bedford*, and contributed to this volume by William H. Tripp, Curator of the Old Dartmouth Historical Society and Whaling Museum. It is represented as being told by Mr. Simmons, the mate.

We was cruising down the Mozambique Channel under reefed tops'ls and the wind blowin' more'n half a gale, two years outer New Bedford an' no ile. An' the masthead lookout shouts, "Thar she blows!"

An' I goes aft.

"Cap'n Simmons," sez I (his being the same name as mine, but no kith or kin, thank God!) "the man at the masthead sez, 'Thar she blows!' Shall I lower?"

"Mr. Simmons," sez the cap'n, "it's blowin' a little too peart an' I don't see fittin' fer to lower."

An' I goes forrard.

An' the man at the masthead sings out, "Thar she blows an' breaches!"

An' I goes aft.

"Cap'n Simmons," sez I, "the lookout at the masthead sez, 'Thar she blows an' breaches!' Shall I lower?"

"Mr. Simmons," sez the cap'n, "it's blowin too peart an' I don't see fittin' fer to lower."

An' I goes forrard.

An' the lookout at the masthead sings out, "Thar she blows an' breaches, an' sparm at that!"

An' I goes aft.

"Cap'n Simmons," sez I, "the lookout sez, 'Thar she blows an' breaches, an' sparm at that!' Shall I lower?"

"Mr. Simmons," sez he, "it's blowin' too peart an' I don't see fittin' fer to lower, but if so be you sees fittin' fer to lower, Mr. Simmons, why lower, an' be good an' God damned to ye."

An' I lowers an' goes on the whale, an' when I comes within seventy-five foot of her I sez, "Put me jest three seas nearer, fer I'm hell with the long harpoon." An' I darted the iron an' it tuk.

When I comes alongside the ship, Cap'n Simmons stands in the gangway. "Mr. Simmons," sez he, "you are the finest mate that ever sailed in this ship. Below in the locker on the port side there's rum and seegars at your sarvice."

"Cap'n Simmons," sez I, "I don't want your rum, no more your seegars. All I want of you, Cap'n Simmons, is plain seevility, an' that of the commonest, God damndest kind."

An' I goes forrard.

Thick. Of the weather, foggy or snowing. Many variant phrases are current at sea and alongshore—dungeon-thick, thick o' fog, or snow, thick as burgoo. To get foggy or snowy is to come up thick. See **Fog.**

Tholes. Wooden pins used as rowlocks. The oars working in them often set up an agonized squeaking—see **Suffer.** In spite of the similarity to the old English word thole, meaning to suffer, endure, the probability is that a boat's thole-pins go back to an Anglo-Saxon word meaning a peg.

Three-piecee bamboo. A three-masted vessel. Pidgin English brought home by sailors.

Three sheets in the wind. See **Sheet.**

Three steps and overboard. A fisherman's walk—the description of a vessel or boat with a very small cockpit. Alongshore, the phrase is used to describe any confined space—for example, a porch.

Thwarts. The seats for rowers, running from side to side of a small boat. According to Weekley, this word comes from thofts, an old English word related to athwart, q.v.

Tide. The sailor distinguishes between tides, of which he is conscious only when in port, and current. To 'longshore people, tide is all-important.

Special terms denote the relation between tide and wind; a leeward ebb means wind and tide both setting out; a leeward flood, both setting in. A windward ebb means the tide going out with the wind blowing onshore, and a windward flood, the wind blowing offshore against an incoming tide. Under these latter conditions, a vessel at anchor is said to be wind-rode or tide-rode, according as she swings with the wind or with the tide.

The exceptionally high and low tides that come once a month are spring and neap tides respectively. A vessel going ashore on the spring tide is said to be neaped or beneaped, since she may have to stay **hard** and fast till the next month's spring tide. Late on the tide, alongshore, covers every possible belatedness.

The word figures in shore speech, from Shakespeare's "there is a tide in the affairs of men," to such phrases as to tide over; the turn of the tide, for the better being usually meant; against the tide or stemming the tide, meaning with great difficulty, contrary to general trend; with the tide, signifying without exertion, or conforming to public opinion. See **Double-tides, Ebb, Flood, Water.**

Tide rips. Choppy waves, occurring on rocky bottom where opposing tides meet. See **Horse market, Merry men.**

Tidewaiter. Originally meant a customs officer who met ships coming up on the tide. In shore speech, it is sometimes figuratively confused with a "time-server," one who is yielding on matters of principles.

Tidewalker. A water-soaked log with only one end out of water; hence, a drifter, a ne'er-do-well. The steamship man's name for it is "propeller-inspector."

Tiller. A steering-bar attached to the rudderpost, formerly used on all vessels, but now replaced by the **wheel** on all but small craft. It meant originally a weaver's beam, but is "naut. from 17th cent." according to Weekley. Alongshore, at the tiller is used interchangeably with at the helm (or wheel). See **Steer.**

Timbered. Constructed. Unfinished wood is used only for **dunnage** aboard ship, and the word lumber is reserved for cargoes. Timber is sometimes used figuratively: "She's well-timbered"; "He wore a timber leg." (The old seaman's timber-toes is now obsolete, as well as his exclamation "Shiver my timbers!")

Time. Time is marked at sea by striking the ship's bell every half-hour during each of the seven watches of the day. At noon, the helmsman strikes eight bells, at 12:30, one bell, and so on through the afternoon watch, adding one bell each half-hour, until 4 p.m. when eight bells is sounded again. The process is repeated from 4 to 8 p.m. (covering the first and second dog-watch); from 8 p.m. to midnight (the first watch); from midnight to 4 a.m. (the middle or gravy-eye watch); from 4 to 8 a.m. (the morning watch); and from 8 a.m. to noon (the forenoon watch). Each bell is repeated by the lookout on the forward bell. In addition, a single stroke is rung on the bell as a warning, ten minutes before each watch changes. It is not uncommon alongshore to hear time alluded to, facetiously, in terms of bells. See **Lights.**

Tip. As applied to the motion of a vessel, this is purely a land-lubber's word. In sailor speech, a ship may cant, capsize, careen, heel or list. See also **Rock** (2).

To'gannels'ls. Pronunciation of topgallant sails.

Togs. Clothes; included in many lists as being a seaman's word. Togged out to the nines—very much dressed up—is supposed to refer to the old-time naval vessel's heavy guns, called long nines.

There is considerable doubt about all this. Togs was English shore slang before 1800; the first sea writer to use it (as toggery) was Marryat in 1837. Togged up to the nines was used in the

London Magazine in 1820, not referring to the sea. Nine is a "mystic number," expressing perfection, which appears in non-nautical folk-phrases. See **Rig** (2).

Tom Cox's traverse. A phrase, current at sea and alongshore to describe sabotage or skillful loafing on the job, is "working Tom Cox's traverse—three turns round the long boat and a pull at the **scuttle butt**." Sometimes Tom goes "up one hatch and down the other." The story, lost in the mists of time, is probably of English naval origin.

Top hamper. Upper yards and rigging. Alongshore, it may mean a large hat, or the upper part of almost anything, especially if heavy or cumbrous. "I think that new hat gives you too much top hamper."

Toplofty. Having tall masts and heavy spars. The word is used in shore speech to mean haughty.

Topmaul. A heavy wooden sledge used in building and rigging ships. Alongshore, a person having difficulty in a job of work will be cheerfully advised to "take a topmaul to it!"

Topping lift (Pronounced top'm'lift). A rope holding a spar or boom in position. It has been used jocosely alongshore for suspenders: "Better take a hitch in your top'm'lift; your pants is draggin'."

Topside. Up. A pidgin English term brought home from China by sailors. It appears occasionally in shore speech; a Washington commentator wrote recently of the topside personnel in a government department.

Tornado. A violent cyclical storm, short in duration and small in area covered. The word is a seamen's perversion of the Spanish tronada, thunder.

Touch at. Of a vessel, to stop at a port of call for a definite purpose, such as to get mail or orders, pick up fresh provisions, or land a passenger. For the distinction from a forced stop when in distress, see **Put.**

Tow. To drag through the water at the end of a rope, as a vessel or barge by a towboat. (Towboat is now used interchangeably with tugboat, though originally they were two different types of harbor craft.)

Alongshore, and to a lesser degree in shore speech, to have, get or take (a person) in tow means to accompany and show him around, to assume an attitude of sponsorship or protection, or to fascinate and enamor. "The way George moons around since that little Jones girl got him in tow!"

Trades, the. The northwest and southeast trade winds, belts of steady winds and good weather found north and south of the equator.

Trade-wind sky. Clear overhead, with fleecy clouds near the horizon. See **Flying-fish weather.**

Trawl-keg (Often pronounced kag, following the original spelling cagge). The buoy at each end of a fisherman's trawl. Alongshore, "Since I was the bigness of a trawl-kag" means since I was a child.

Trice, Trice up. To lift, suspend. Specifically, on old warships, to spread-eagle a man preparatory to flogging. Alongshore, it is used to mean to raise, as a window blind, or a woman's skirt (when they were worn long).

Trick. A turn at the wheel or on lookout. Alongshore it means a turn at doing something. "It's your trick at the dishpan tonight."

Trim. (1) Of sails, to adjust them so as to get the full benefit of the wind.

(2) To distribute cargo so as to keep the ship on even **keel.** When thus loaded, she was in good trim—a phrase current in shore speech to mean in good order or health. In fighting trim comes from naval warfare.

Alongshore, to trim means to distribute weight evenly. "You two change places, so the hammock will trim better." See Sail.

Trip. A short **voyage.** A term borrowed by the railroads—as single trip, round trip. Alongshore, a "trip o' fish" is the fisherman's catch on a single voyage. See **Fare** (1), **Passage.**

Truck to keelson, from. The truck is the extreme tip of the mast; the keelson (pronounced kilson) is one of the timbers of the keel. Hence, at sea and alongshore, the phrase means from top to bottom; all over. See **Alow** and aloft, from **clew to earing,** from **stem to sternpost.**

True blue. Loyal, staunch—a sailor's term, referring either to the color of the deep sea, or to that of a naval uniform.

Tug. Short for tugboat. See **Tow.**

Tumble home. To slope inward from the bottom up, as a vessel's topsides. Used in the same sense alongshore: "You build a sea chest with the sides tumbling home, so's it will set steadier." See **Home.**

Tumble up, or out. The call to the watch below to come on deck. Alongshore, it is used to waken sleepers. See **Show a leg, Turn out.**

Turn. Of a rope, one loop thrown round a pin and held. Alongshore, to hold the turn is to play a very small part in an event. See **Bring up, Hitch** (2), **Slack.**

Turn in. To go below at the end of one's watch. To turn in all standing is to sleep without removing one's clothes. To turn in, meaning go to bed, is in common use ashore.

Turn out; Turn to. To come on deck after one's watch below; to go on duty. See **Knock off.** Alongshore, to turn out is to get out of bed; to turn to is to go to work, get busy. "I've got to turn to and finish mowing the lawn."

Turtle. The English name for the tortoise (French *tortue*, Spanish *tortuga*) is believed by etymologists to have been bestowed by sailors, and to have been influenced by *turkey*, because of the long stringy neck common to both creatures. "Turkle" is noted as being a dialect variant of turtle.

To turn turtle—that is, **capsize**—probably comes from the sailor's observation of the fact that a turtle placed on its back

is unable to turn over. The phrase is in common use among landspeople, about other objects than ships.

Twice-laid. Rope twisted from old rope yarns is said to be twice-laid. See **Lay** (2). The term is applied alongshore to re-cooked leftovers, made over garments, and so on.

Twist, Twister. A tall story, a lie. See **Yarn.**

Two cents and a fishhook. A nautical phrase indicating a small consideration: "I wouldn't give two cents and a fishhook for it."

Two lamps burning and no ship at sea! A 'longshore reproach to someone who is extravagant or wasteful.

Typhoon. Tai-fung, Chinese for the cyclical storms experienced in the China Sea and islands of the Pacific; a term brought home by sailors.

Uncharted. Not surveyed; not appearing on the chart. It is used figuratively ashore to mean unknown, hazardous, or remote; uncharted ways (or dangers), the voyager upon uncharted seas, etc.

Underwrite. An insurance term, meaning to share risks among subscribers called underwriters. The first insurance was applied to vessels at sea. In addition to its still current use by insurance companies, the term is used by shore people in a general sense of agreeing in advance to make up deficits that may be incurred in launching some new project.

Union down. In distress; referring to the practice of flying a ship's ensign upside down, which brings the Union Jack at the bottom instead of the top. See **Colors.**

Unship. See **Ship** (5). To detach from a prepared setting, as oars, or a rudder, and said alongshore as well of wheels, tires, and so on.

Vamoose. "Get out of here!" A Spanish word, probably brought home to the coast by sailors (as in the case of **pronto,** q.v.).

Vast. See **Avast.**

Veer. Of the wind, to **haul;** the opposite of **backen.** Veer is never used at sea or alongshore in its dictionary sense of to change direction, except when speaking of the wind. To veer out is the same as to pay out.

Vessel. Any craft larger than a **boat**—to all intents and purposes the same as **ship** in its generic sense. A common plural is **sail** o' vessels. "Three sail o' vessels were in sight."

Vessel property. Shares owned in vessels, reckoned in halves, quarters, eighths, sixteenths, thirty-seconds, or sixty-fourths of the value. In the cooperative building and ownership of wooden vessels on the coast, the master was expected to own (or control through his friends and relatives) a substantial amount of the ship he commanded, and buying into a vessel was understood to mean acquiring a master's interest. "Did you hear tell that Cap'n Jim Harriman is buyin' into the *Resolute?*" conveyed also the news that Cap'n Jim was about to take command of her.

Voyage. This word originally meant travel of any description; later it was restricted to travel by sea. Weekley calls this "a limitation of sense characteristic of a seafaring nation [England]." To the modern sailor, it means to go out and home. See **Trip, Passage. A deepwater** voyage under sail was seldom less than a year in length; sometimes it might last for two or three. Long voyages were a commonplace to whaling men. Reminded that he was sailing without having said goodbye to his wife, one of them is said to have replied, "Never mind, I won't be gone more'n a couple o' years."

A vessel's first voyage is her maiden voyage. This term is used figuratively alongshore, together with many other vigorous sayings using the word. All in the course of the voyage is the same as all in the day's work. It takes a voyage to learn is the excuse for someone who has bungled an unfamiliar task. To last out the voyage means to be sufficient for a given period, or for one's lifetime.

Voyage appears in literary phrases ashore—the voyage of life, voyager upon the sea of destiny, and the like.

Wagon. A sailor's cant name for a ship, particularly one that is slow and unwieldy.

Wake. From the old Norse word for a passage cut in the ice through which a ship might pass. In modern sea speech, it is the track which a ship leaves behind her in the water. A good helmsman makes a straight wake. This phrase is used alongshore to mean go directly, without delay. Of a tricky person, as of a bad helmsman at sea, it is said "It would break a snake's back to follow his wake." In shore speech, in the wake of a person or event means behind, following or ensuing.

Walk away. When the wind helped the process of swinging the yards in tacking ship, the slack of the braces came in so easily that the crew had to walk or run down the deck to keep the rope taut. In shore speech, to walk away with something is to experience no difficulty. (Not to be confused with win at a walk, which comes from horse racing.)

Walk the plank, to. Pirates of old are said to have disposed of unwanted prisoners by placing the gangplank with its end over the bulwarks, and forcing them to walk overboard and drown. The phrase is occasionally heard in shore speech, concerning someone dismissed in an inconsiderate manner, or made a scapegoat.

Watch. (1) The crew is divided into watches, usually two in number, the starboard watch being called the captain's watch, and officered by the second mate, and the port watch being headed by the mate.

(2) Watches are also the periods during which the crew are on duty (stand watch), usually four hours on, the watch on **deck;** the four hours off being the watch below. See **Time.**

Alongshore, to call the watch means to summon to duty. "Go below the watch," the conventional term of dismissal at sea, is used facetiously in dismissal alongshore after a job is finished. To stand (or keep) watch and watch is to share the job of watching.

"Sam and Bill sat up with Henry watch and watch when he had pneumonia." To keep watch of (or over) something does not come from the sea, but from watchmen ashore—sailors stand watch. See **Anchor watch, Battle the watch, Dogwatch, Everybody's mess. Watch tackle.** See **Jig, Tackle.**

Water(s). (1) Depth of water under the keel. Its sea uses are represented in shore speech chiefly by the phrases in clear (smooth) water, signifying uncomplicated; and in rough water, or in difficulty. See **Back water,** plenty of water under the **keel.**

(2) In certain phrases, water equals **tide**—high water, low (dead low) water, slack water, etc. See **Hell and high water, High-water mark.**

(3) Waters, referring to the sea, is purely a landsman's word, used in proverbs and literature. "The waters of the earth—"; sail o'er troubled waters; still waters; etc. See **Oil** on troubled waters, **Fish** in troubled waters. (In deep waters, meaning in trouble or sorrow, and to keep above water—manage with difficulty—probably come from swimming rather than from navigation. Deep waters constituted no hazard to the seaman.)

Water bewitched. The sailor's term for a very weak mixture of spirits and water. Alongshore, it is applied to weak tea, coffee, etc. On Cape Cod, thin porridge was described as "water bewitched, meal begritched" (begrudged).

Water front. That part of a port where the docks lie. Probably a sailor's term, originally.

Waterlogged. Filled with water but still afloat. See **Derelict, Founder, Swamp.** Ashore it means having drunk too much liquid; being all afloat inside. It may also mean logy, difficult to control.

Wave. This word does not appear in English until late in the 16th century. It is seldom used nowadays in the seaman's vocabulary, but is purely literary—*The Wave of the Future;* on top of the wave; o'er the wild waves; the sport of wind and wave; etc. The seaman's terms are **sea** and **breakers.**

Way. Of a ship or boat, progress through the water. A vessel gets under way (never weigh; that is a landsman's error) by weighing anchor and setting the sails. She then gathers way; is under way. This phrase is in common use ashore to mean initiating a project: "The construction of the plant is well under way." See **Away, Aweigh.**

In a boat, the order to begin rowing is "Give way!"; to stop is "Way enough!" The latter phrase is used alongshore to mean stop, or don't drive any further. To give way, meaning to break under strain, may be of sea origin.

Ways. The track down which a vessel slides at launching. To grease the ways, meaning to facilitate, is common alongshore and is also found occasionally in shore speech. Off the ways is sometimes used to mean completed, ready for use.

Weather. (1) To get to **windward** (of). To weather a storm is to come through it safely, and is used figuratively in precisely this sense in shore speech: "I think we'll weather it this time."

(2) Besides its ordinary significance, the seaman combines the word weather with make, to indicate the vessel's performance. She makes heavy weather or better weather of it. These phrases have been adopted into shore speech to mean getting along with difficulty, or more easily. For under press of weather, see **Weather-bound.**

(3) As an adjective, weather is the opposite of **lee;** toward the windward side, but usually within the ship—the weather fore-braces, etc. The word enters many shore phrases—under the weather means slightly indisposed; from the condition of the seasick **green-horn,** who sought the shelter of the weather bulwarks.

To have a weather(ly) eye, or to keep a weather eye lifting has the same significance as an eye to **windward.** To get the weather gage of someone, meaning to secure a decisive advantage, comes from naval warfare under sail, when the vessel which succeeded in getting to windward of her adversary was in a favorable position to attack.

Weather-bound. Of a ship, kept in port and prevented from sailing by adverse weather conditions. Alongshore, it means kept indoors—by a heavy snowfall, etc.

Weather-breeder, a. A day of unusual stillness and clearness, said to forerun a storm.

Weft. To signal with a flag; to wigwag; a practice on whaling vessels, when signaling to boats in pursuit of a whale. Common alongshore to refer to similar signals.

Well. See **Forms of address.**

West Coast, the. This means of South America or of Africa, as implied by the context. The corresponding region of North America is the Pacific Coast, or merely the **Coast.**

Westerlies, the. A belt of prevailing westerly winds south of Cape of Good Hope. See running (one's) easting down.

Western Islands, the. The Azores.

Western Ocean, the. The North Atlantic. A very old term, brought over from England, and still in use, although to us, the Western Ocean lies to the eastward.

Westing. Progress to the west; a vessel makes westing.

West Injies, the. Pronunciation of West Indies.

Westward (Pronounced west'ard). Toward the west.

Whale of a. A whaleman's term, indicating something very large or important. It is common alongshore, and to some extent in shore speech. See also a **dead whale.**

Whale, to heave a tub to a. To throw a sop to someone, or attempt to distract his attention; from the old whaling custom of giving a frightened or ugly whale something beside the boat on which it could vent its rage. Swift, in his *Tale of a Tub*, speaks of this custom.

Wharfinger (pronounced with soft "g"). The owner or person in charge of a wharf.

Wharf rat. A sneak thief operating along the water front.

Wheel. A modern device for controlling a ship's rudder. See **Helm, Man** at the wheel, **Tiller.**

Whip. (1) A rope rove through a single block, used for hoisting light objects. Used in the same sense alongshore.

(2) At sea and alongshore, to whip is to wrap the end of a rope with twine to prevent fraying. Seamstresses whip, that is, overcast or oversew, the raw edges of a seam for the same purpose. This may come from the sea use of the term (or the other way round).

White-ash breeze. The current of air created when a boat is rowed in a dead calm. Ash is the best wood for oars.

White slaver. A person who procures and transports women for purposes of prostitution. Reference is to the people and ships that brought black slaves from Africa.

Whole-sail. See **Breeze** (1).

Whopper, a. Something enormous or impressive; specifically, a lie. Weekley says it is a landsman's word, imitative of the sound of slapping; Smyth claims it for the sea, and says sailors made it up from the name of the huge turtles called guapas which are found in the West Indies—an ingenious theory, anyway, but it may be a "whopper."

Winch (Pronounced wench). A mechanical device used on shipboard, consisting of a barrel turned with a crank, for hoisting and winding; a windlass. The word is used in connection with similar operations alongshore.

Wind. (1) A chief preoccupation of the sailing-ship man. At sea and alongshore, many terms are used to express its force and direction. See **Air, Backen, Blow, Breeze, Calm, Catspaw, Gale, Haul, Storm, Squall, Veer.**

Of wind and rain the seaman sang:

First the rain and then the wind,
Topsail sheets and halliards mind;
First the wind and then the rain,
Hoist your topsails up again.

A capful o' wind is a nice little sailing breeze. Directly into the wind is in the eye (or teeth) of the wind, or the wind's eye.

Many phrases current in shore speech go back to the sailor's experience with wind. It's an ill wind that blows nobody any good means that any wind must be fair for some ship. Hanging in the wind, meaning uncertain, dependent upon chance, comes from the situation of a vessel in stays.

To see how the wind blows, which has the same figurative meaning as see how the land lies, as well as the exclamation, "So that's how the wind blows!" on discovering facts hitherto unsuspected, are of nautical origin. So also is the slang phrase, to raise the wind, meaning to secure money or credit. It comes from the seaman's practice of whistling for the wind during a calm. This latter phrase has itself entered shore speech, to mean talking to keep up one's own, or others', courage and spirits. (To get wind of and there's something in the wind are, however, from hunting and not from seafaring.) See trim one's sails (1).

(2) In relation to navigation, fair (favoring) winds include:

A following wind, one astern or from abaft the **quarter** (see also **Wing** and wing).

A quartering wind (or wind on the quarter), coming roughly from an angle of 45° to the course.

A beam wind (wind abeam, on the beam), or at approximately right angles to the course. (Up to this point, the vessel is said to be "sailing large.")

A leading wind, one from forward of the beam, but still permitting the ship to be steered by compass.

A head wind is any wind forward of a leading wind, which necessitates sailing off the wind, full and by, on, **nigh,** or **by the wind,** or **close-hauled.** See also **Beat, Windward.** The exact point between a leading wind and a head wind varies with the vessel, depending upon how well she is able to point up into the wind.

All these phrases are used, literally and figuratively, in coastal dialect; but shore speech knows only (figuratively) fair (favoring) winds, signifying propitious circumstances; and head winds, or adversities.

In relation to the land, winds are on-shore and offshore. For their relation to tides, also for wind-rode, see **Tide.** See also **Run** (2), **Scud;** busy as the **Devil** in a gale of wind; **Sail** close to the wind; take the wind out of one's **sails;** three **sheets** in the wind.

Wind-bound. Kept in port by a head wind. See **Weatherbound.**

Windjammer. A disparaging name for a sailing vessel, invented by steamship men. Shore people often use it to demonstrate their familiarity with sea terms, without realizing how it grates on the ear of a man brought up in sail.

The sailing-ship man's retorts to windjammer were humbug and derricks, kettle and coffee mill, and smoke box. To these the Banks fisherfolk added "Silent Death," from the way steamers rampage out of the fog to cut down anchored fishermen. See also **Go into steam.**

Windlass-bitts. The post to which the end of the anchor chain is attached. "Between you and me and the windlass-bitts" is common alongshore—the equivalent of "you and me and the bedpost" in shore speech.

Windward (pronounced wind'ard). The opposite of **leeward;** intensified to **dead** to windward. Toward the direction from which the wind is blowing, but outside the ship. The distinction between windward and weather can best be shown by an example: "He stood by the weather bulwarks, watching the brig to windward." See **Weather** (3).

Alongshore, to claw, eat or work to wind'ard, vividly expresses the difficulty of beating against a headwind, and means, figuratively, to work oneself gradually out of a tight spot. "I'm clawin' to wind'ard of that debt little by little."

The young seafarer is enjoined: "Don't try to spit to wind'ard"—equivalent to the landsman's advice to keep one's nose clean. The greenhorn may also be instructed never to heave over to windward anything except hot water and ashes from the galley—and then be left to learn from experience. To go ashore to windward is a description of extreme ineptness or incapacity.

Landsmen occasionally use the phrase to get to windward of someone, meaning to gain an advantage—usually by unfair means. See get the **weather** gage. (A 'longshore speaker, would be more likely to cut in to wind'ard.) To have an eye to windward denotes prudence, in shore speech as well as alongshore. See also **Anchor** to windward, **Beat.** For windward ebb and windward flood, see **Tide.**

Wing and wing. Said of a schooner when running before the wind, with booms out on either side. In this situation she is said to be winged (or preferably, wung) out. The term is occasionally used figuratively alongshore to mean proceeding at a fast pace.

Woold (Dutch, woelen). To strengthen, as a spar, by wrapping rope round and round it.

Work. To become loosened, to creak and complain, as a wooden ship under stress of sea and wind. Used in the same sense alongshore of something rickety.

Workaway. A word manufactured in imitation of stowaway

meaning a person who volunteers to exchange work for his passage. It has not appeared in shore speech.

Work (one's) old iron up, to. A phrase current at sea and alongshore, meaning to punish or discipline severely. (Sometimes shortened to "work up.") It probably refers to the hated task of chipping the rust off ironwork on board ship, preparatory to painting it.

A sailor's rhyme, which somehow or other has acquired the name of the "Philadelphia Catechism," runs as follows:

On six days shalt thou labor and do all that thou
    art able,
And on the seventh holystone the deck and scrape
    the cable.

Work one's passage. See **Passage.**

Wreck. (1) The remains of a lost ship. A total wreck is one upon which full insurance is claimed. This phrase is current ashore, in such expressions as to look, feel, like a total wreck. See **Shipwreck,** total **loss.**

(2) To wreck a vessel is to cast her away and lose her. Ashore, to demolish, raze, as a building—with many derivative words, such as wrecking companies, wreckers, "wreckaroos."

Wreckage. Fragments of a wrecked ship or her cargo. The word is used figuratively in shore speech—the wreckage of hopes, plans, and so on.

Wrong tack. See **Tack.**

Yard. Spar from which the head of a sail is suspended. See **Boom.**

Yardarm. The tip of a yard. An officers' quip, that comes from the old sailing navy, is "When the sun is over the yard-arm (or sometimes the fore-yard or fore-yardarm) it's time to take a drink." In some occult way, this means that it is noon, before which it was not etiquette to order drinks in the wardroom. In modern shore speech, yachtsmen, and even landlubbers, may be heard anxiously inquiring of each other whether "the sun is over the yardarm." See to correct the **declination.**

Yarn, to spin a. In making spunyarn from untwisted yarns of rope, it took two sailors to operate the small winch. They worked in some sheltered spot, enlivening their task with conversation. Hence, to spin a yarn, or a twister, is to tell a story, usually a "tall" one. The phrase has passed into a verb, to yarn, which means to chat aimlessly and interminably. It is quite common in shore speech.

Yaw. To steer wildly, be almost unmanageable; used in the same sense of vehicles in coastal speech.

Yellow Jack. The sailor's name for yellow fever. It comes from the **quarantine** flag (the code flag Q, yellow in color), which is hoisted when there is infectious disease on board. The term is understood and used by landsmen—was, in fact, the title of a successful play.

Yeo-heave-ho. The standard literary spelling of those "unnameable and unearthly howls" which sailors emit when singing out on a rope. A poem of the 14th century has the lines:

> Your maryners shal synge arowe
> Hey how and rumby lowe.

    With these ancient but well-remembered tones ringing in our ears, it is fitting to bring this labor of love to a

FINIS

"DOUBTFULS"

Shore expressions included in some lists of nautical terms, but upon whose sea origin doubt has been cast. References to word list are indicated as throughout text.

All **set**
Betwixt the Devil and the deep blue **sea**
Bowers (in euchre)
Bowl along
Brace of shakes
Bully (meaning fine)
Caboose (jail)
Cutwater (of a bridge)
Company, in; keep; part
Crank (a monomaniac)
Don't lose the ship for a pennyworth of **tar**
Fly the blue pigeon
Gaff, blow the; stow your
Greenhorn
Keep **watch**
Kick the **bucket**
Land (a job, etc.)
Larrup
Leave in the **lurch**
Let **fly**
Lick of the **tarbrush**
Line out; see **Lines**

Line up
Lubber; see **Lubberland**
Maneuver
Manhandle
Martingale
Not by a long **chalk**
On the **square**
Passage; Passenger
Play **fast and loose**
Put on **lugs**
Quarantine
Roped in; see **Rope**
Skyscraper
Square off
Stick to one's **colors**
Tackle (meaning to attack)
Tally
Tarred with the same brush
Till all's **blue**
Togged out to the nines; see **Togs**
Under color of; see **Colors**
Whip (oversew)
Whopper